COLLOIDS

BY

ERNEST S. HEDGES

M.Sc., Ph.D. (Manchester), D.Sc. (London), A.I.C.
LATE DARBISHIRE FELLOW OF THE UNIVERSITY OF MANCHESTER

LONDON

EDWARD ARNOLD & CO

1931

All rights reserved

Made and Printed in Great Britain by
Butler & Tanner Ltd., Frome and London

PREFACE

To some, the word " colloidal " conjures up visions of things indefinite in shape, indefinite in chemical composition and physical properties, fickle in chemical deportment, things unfiltrable and generally unmanageable. What could be worse ? Actually, that is a very mistaken idea of the study of colloids. In the first place, the colloidal state is characteristic of all the material things that most nearly concern us—our bodies, our food, and our clothes—and, secondly, the modern developments of colloid chemistry have removed most of the indefiniteness in which it was formerly enshrouded. In fact, there are, perhaps, few branches of physical science where it is possible to get a better mental image of the processes involved. This visual impression is to be gained, not by regarding the colloidal system as the unit, but rather by considering the processes from the point of view of the colloidal particle itself. Only by getting down to the colloidal particle itself in imagination can this vivid picture, which is essential to the proper understanding and appreciation of colloid chemical principles, be obtained. It is from this point of view that I have written the present book.

For some years I delivered courses of lectures on Colloid Chemistry at Bedford College, University of London, and my students repeatedly asked me to write a non-mathematical textbook of colloids. I have adhered to this request as far as possible, and whilst the book has been written more especially for students of chemistry I have endeavoured to treat the matter in a sufficiently broad way to interest students of physics, biology, and geology.

So much for the students' point of view. For some years I have been preparing abstracts of a considerable portion of the current colloid chemical literature and I have incorporated in the book such accounts of this recent work as seemed suitable, not only with the object of making the book as up-to-date as possible, but also with the aim of serving those engaged in research on colloids. Some of these recent lines of investigation have not been compiled in this way before.

In order to serve this double purpose, the references to the literature have been left in the body of the text rather than placed at the end of each chapter. The reader can then see at a glance what is recent research. The references have been so chosen that they refer either to classical papers of the older literature or to developments of special interest during the past few years.

The balance of the book is not quite the same as in many other books on colloids. The properties of gels and the peculiarities of chemical reactions and other processes taking place therein have received greater attention in relation to the rest of the book. This has been done intentionally, because natural colloids appear more often in the form of gels than sols, and biological and other reactions normally take place in gels. It seems a pity, therefore, that this most important aspect of the study of colloids receives on the whole comparatively little attention.

The technical applications of colloid chemistry are so numerous that they would merit a separate volume. No attempt has been made to describe technical processes in the present book, but the last chapter endeavours to show some of the ways in which colloid chemical principles have been applied successfully to industrial and other processes.

E. S. H.

June, 1931.

CONTENTS

CHAP		PAGE
I	GENERAL INTRODUCTION	1

Scope of the study of colloids. Early work on colloids.

| II | COLLOIDS AS POLYPHASE SYSTEMS | 5 |

Homogeneous and heterogeneous systems. Application of the ultramicroscope. Wo. Ostwald's generalization of dispersoid chemistry. Types of disperse systems. Disperse phase and dispersion medium. Degree of dispersion. Von Weimann's dispersion coefficient. Types of colloid systems. Hydrosols. Water as a disperse phase.

| III | THE FORMATION OF COLLOID PARTICLES | 14 |

Preparation of colloid systems. Dispersion methods (a) mechanical dispersion, the colloid mill, (b) peptization, (c) electrodispersion, (d) dissolution of salts. Condensation methods: (a) reduction, preparation of gold sols, Carey Lea's silver sols, (b) oxidation, (c) dissociation, (d) exchange of solvent, (e) double decomposition, (f) hydrolysis. The process of condensation; formation and growth of nuclei.

| IV | GENERAL CHARACTERISTICS OF COLLOID SYSTEMS | 31 |

Osmotic pressure and related phenomena. Molecular weight. Uniformity of particles. Polydispersity and isodispersity. Preparation of isodisperse systems: (a) Perrin's centrifugal method, (b) Zsigmondy's nuclear method, (c) Odén's fractional coagulation method, (d) by ultrafiltration. The Tyndall effect. Scattering of light. Rayleigh's law. Application of the ultramicroscope and its limitations. The Brownian movement. Calculation of the Avogadro constant. Determination of particle size. Shape of the particles. Properties of sols with non-spherical particles. Crystalline nature of the particles. X-ray investigation of the stretching of amorphous substances.

| V | ELECTRICAL PROPERTIES OF COLLOIDS | 51 |

The charge on the particles. Cataphoresis. Charge and stability. Electro-osmosis. Streaming potential. Methods available for determining the charge. The electric double layer; the diffuse double layer. Electrostatic and electrokinetic potentials. The origin of the charge. Mechanism of peptization and coagulation. Constitution of metal hydrosols.

CONTENTS

CHAP.		PAGE
VI	PRACTICAL METHODS OF COLLOID INVESTIGATION . .	70

Special precautions required. Dialysis; electrodialysis. Ultrafiltration. The ultramicroscope. Cataphoresis; macro- and micro-methods. Viscosity.

VII CLASSIFICATION OF COLLOIDS 82

General features. Reversible and irreversible. Lyophobic and lyophilic. Suspensoid and emulsoid. Characteristics of the classes. The " evolution zone."

VIII STABILITY OF COLLOIDS 89

Effect of gravity. Stokes' law. Effect of electrolytes. Coagulation. The Schulze-Hardy rule. Antagonistic effect of the similarly charged ion. Explanation of the valency rule. Probability of collision and probability of adhesion. Velocity of coagulation. Effect of concentration of the sol. Adherence. Acclimatization. Irregular series. Interaction of two hydrophobic colloids. Coagulation by radiations, heat, and mechanical agitation. Peptizability of a precipitate. Interaction of hydrophobic and hydrophilic sols. Sensitization, U-numbers. Protection, gold numbers, theory of protective action, preparation of protected sols. Supersaturation in the presence of colloids. Ageing of sols.

IX ADSORPTION 119

Adsorption and absorption. Specific surface. The liquid-vapour boundary. Surface tension. The liquid-liquid boundary. The solid-gas boundary. The unimolecular layer. The liquid-solid boundary. Chemical phenomena at boundaries. Positive and negative adsorption. The adsorption isotherm. Influence of the solvent. Specific effects. Exchange adsorption. Lyosorption. Electroadsorption. Adsorption of electrolytes. Rôle of adsorption in coagulation and peptization. The Langmuir-Harkins theory. Solubility.

X PROPERTIES OF HYDROPHOBIC COLLOIDS . . . 144

Stability. Surface tension. Viscosity.

XI EMULSIONS 149

General properties. Phase reversal. Influence of phase-ratio. Influence of the emulsifier.

XII PROPERTIES OF HYDROPHILIC COLLOIDS . . . 154

Difficulties of investigation. Viscosity. Poiseuille's law. The electro-viscous effect. Surface tension. Foaming. Stability towards heat and electrolytes. Kruyt's theory of simultaneous discharge and dehydration. Salting-out. The Hofmeister or lyotropic series. Phenomena at the isoelectric point.

XIII INDIVIDUAL HYDROPHILIC COLLOIDS 168

General properties of proteins. Gelatin. Albumins; denaturation. Globulins. Hæmoglobin. Casein. Molecular weight of proteins. Hydrosols of carbohydrates. Starch; starch iodide. Cellulose; artificial silk. Agar. Soaps; detergent action. Silicic acid.

CONTENTS

CHAP.		PAGE
XIV	SOME NON-AQUEOUS COLLOIDS	190

Organosols and organogels. Mercury as a dispersion medium. Pyrosols.

XV THE PROPERTIES OF GELS 195

Gels and jellies. Setting. Hysteresis. Thixotropy. Swelling; swelling pressure, volume and heat changes. Dispersion in salt solutions. Shrinkage. Muscular contraction. Syneresis. Mechanical and optical properties. Effect of stretching. Structure of gels. Classification. The solid-phase rule. Thermodynamics.

XVI DIFFUSION AND CHEMICAL REACTION IN GELS . . 219

Diffusion; methods of measurement, equilibrium between colloid and crystalloid, Pringsheim's rule. Crystallization and precipitation; particle size. Chemical reaction. Precipitation structures. The Liesegang phenomenon. Theories of periodic structures. Distinction between periodic structures and periodic reactions. Periodic structures formed by coagulation and by salting-out. Natural periodic structures.

XVII COLLOIDS AND CHEMICAL REACTIVITY . . . 241

Direct reactions of colloids. Velocity of reaction. Catalysis Retarding effect of hydrophilic colloids. Periodic reactions.

XVIII SOME APPLICATIONS OF COLLOID PRINCIPLES . . 247

Smokes and fumes. Clays and soils Colloidal graphite. Qualitative and quantitative analysis. Dyestuffs. Leather Rubber Milk. Wool Water purification Sewage disposal. Photography.

AUTHOR INDEX . 265

SUBJECT INDEX 268

COLLOIDS

CHAPTER I
GENERAL INTRODUCTION

In a certain sense the physics and chemistry of colloids form a more fundamental study than the "pure" sciences. It is true that in order to investigate and understand colloid systems a sound knowledge of the principles of pure science is essential; but the study of colloids is more fundamental in the sense that it is more true to Nature, and Science is the systematized study of Nature.

This point of view is particularly well brought out in the chemistry of colloids. "Pure" chemistry is an *idealized* chemistry, which the chemist studies by working only with highly purified substances, having quite definite and universally reproducible physical properties. The pure chemist may go even further and refuse to examine substances except those having a constant chemical composition—the so-called chemical compounds. Other substances of variable composition he may regard as mixtures of the constant compounds which form his study, and although he may study each "chemical" component of that mixture he will not be interested generally in the mixture itself.

A bird's nest may be made of twigs, straw, wool, and horse-hair. The pure chemist is represented by an investigator who makes a thorough study of twigs, straw, wool, or horse-hair, or all of these substances, but who may not be interested in the most important natural object of his study—that it is a bird's nest. A heap of twigs, straw, wool, and horse-hair does not make a nest, because a definite structure is involved, and it is this combination of the study of structure as well as chemical composition that forms one of the chief differences between the outlooks of pure chemistry and colloid chemistry.

Colloid chemistry is a closer approach to the study of things as they exist. Purity in Nature is the exception, not the rule. In the laboratory we may purify a natural product until eventually we obtain something having a chemical composition, crystalline form, and physical properties which are constant and reproducible ; but this idealized substance may be quite useless, whilst the natural product (e.g. wool or cotton) may have been used for centuries in virtue of the properties which are dependent not only on its chemical composition, but also on its structure. The study of colloids treats of these substances in their impure, natural state.

The chemist works with such ideal substances as alcohol, sodium chloride, etc., and throws away muds, slimes, tars, and emulsions. Although these discarded substances come into the study of colloids, it would not be right to suppose that colloid chemistry is typified by so unwelcome an assembly. Colloids enter into all the material affairs that most nearly concern us.

In the first place, our bodies are composed almost entirely of matter in the colloidal state; our food, consisting mostly of other organisms, is also mainly colloidal ; our clothes are made of typical colloidal substances ; and colloids form a large part of our property. So that a knowledge of the phenomena associated with colloids is essential in biology and physiology ; the study of colloids is important in dealing with foodstuffs and also in the administration of medicines, many of which are given in the form of ointments or emulsions—typical colloid systems ; the production of fabrics of silk, wool, cotton, linen, leather, and especially artificial silk are problems of colloid chemistry ; and colloids enter into many fundamental branches of industry, such as agriculture (where the colloidal constituents of the soil are of importance), sewage disposal, and the production of rubber, paper, dyestuffs, pottery, glassware and a large number of other substances.

It will be clear that colloids must have been studied in an empirical way from the earliest times, but the systematic study of colloids is of comparatively recent date, and indeed the physics and chemistry of colloids have been developed mostly during the present century. The science is therefore a rapidly growing

GENERAL INTRODUCTION

one, and one which promises to attain even greater importance and interest in the future.

It is generally agreed that the systematic study of colloids began with the researches of Thomas Graham in 1861 (*Phil. Trans.*, 1861, **151**, 183). It should be mentioned that substances now recognized as colloids had previously been prepared by artificial means and their properties had been investigated, but not in a systematic way. For example, colloidal gold, in the form of " Purple of Cassius," was known as early as 1685 ; Berzelius, and later Wackenroder, in the first half of the nineteenth century prepared and examined colloidal solutions of sulphur, and in 1857 Faraday (*Phil. Trans.*, 1857, **147**, 145) prepared colloidal solutions of gold by reducing extremely dilute solutions of auric chloride with ethereal solutions of phosphorus. Faraday noted many of the properties of these purple solutions and with characteristic insight expressed the view that they contained metallic gold suspended in an extremely fine state of subdivision.

Graham's classical researches were on the relative rates of diffusion of dissolved substances. For this purpose the solution was separated from the pure solvent by means of a membrane such as parchment, and the rate of decrease of concentration of the solution gave a measure of the rate at which the solute diffused away through the membrane. This process is called dialysis and its practical aspect is discussed in Chapter VI. The membrane through which the solute diffuses must not be confused with a *semi-permeable* membrane, which allows the molecules of the solvent to pass, but not those of the solute.

The fundamental discoveries of Graham were two in number. In dialyzing a number of solutions through parchment membranes, he found that some passed readily through the membrane, whilst others diffused very slowly or not at all. The former class of solutions contained substances which can generally be obtained simply in a crystalline form (sugar, sodium chloride, etc.), and the latter class contained substances which are generally not crystalline (gelatin, albumin, glue, etc.). Graham therefore called the former class " crystalloids " and the latter class " colloids " (from κόλλα, glue).

The other fundamental discovery of Graham was that solutions of normally insoluble substances could often be obtained by using special methods and that these solutions belonged to the colloidal class, since they did not diffuse through a membrane of parchment. The special methods involved will be discussed later, and it is necessary only to mention here that colloidal solutions of such substances as silicic acid, ferric hydroxide, and the hydroxides of chromium and aluminium were prepared by Graham. These solutions were apparently clear and homogeneous, but differed from "true" solutions of crystalloids in ways other than their behaviour on dialysis ; for example, on the addition of electrolytes, many of them set to jellies or were coagulated to form a flocculent precipitate. To distinguish these colloidal solutions from "true" solutions, Graham referred to them as "sols," and this term has been retained ; the gelatinous products of coagulation he called "gels."

CHAPTER II

COLLOIDS AS POLYPHASE SYSTEMS

Since the time of Graham our conception of colloids has undergone a radical change. Graham thought of colloids as *substances*; nowadays we apply the term not so much to the substance, but rather to a *state* in which that substance exists. Even Graham realized that behaviour on dialysis did not allow a division of substances into two distinct classes and that there was a gradual transition from one to the other, as indeed is characteristic of almost all classifications. Nevertheless, he employed the terms " crystalloid " and " colloid " to denote the substances themselves. In accordance with newer ideas it is more correct to speak of " colloidal systems," meaning thereby a particular physical system generally consisting of more than one substance. Colloidal gold is a physical system composed of water and of gold in a certain state of subdivision ; there is no reason to suppose that the actual gold particles differ essentially from any other gold. However, it is still quite usual and is often convenient as a matter of expression to refer to the gold itself or other substance as a colloid, and this is especially the case when dealing with substances of the type of gelatin, cellulose, gums, etc., where a crystalline state is not generally encountered.

This complete change of viewpoint has been rendered necessary through the preparation of colloidal solutions of a large number of insoluble substances which are usually known as crystalline bodies. In particular, the last thirty years of research have shown that in water alone thousands of colloidal solutions of normally insoluble metallic hydroxides, sulphides, sulphates, carbonates, phosphates, chlorides, etc., and non-metallic substances such as sulphur, selenium, and graphite can readily be prepared. All these substances either occur naturally or can be obtained by simple means in a crystalline state also. Evidently, the distinction between colloid and crystal is a matter connected not with the substance itself, but with the state in which the substance can exist. It is true that some substances can be obtained in colloidal solution much more readily than others, but it is probable that given the right conditions a colloid system of any known substance could be obtained. Naturally, the solvent or liquid dispersion medium is another important factor and shows that the term colloid cannot logically be applied to the dissolved substance, for many substances dissolve in one solvent to form a typically crystalloidal solution and in another to form a colloid system. Soap is an example of this. A solution of soap in water has all the characteristics associated with colloids, but soap dissolves in alcohol to form a " true " or crystalloidal solution. Therefore, it is not strictly logical to regard soap as the colloid, but rather the aqueous soap solution as a colloid system.

Heterogeneity of Colloid Systems. When viewed by transmitted light, most colloidal solutions appear to be perfectly clear and homogeneous, and in this respect resemble a true solution. When, however, the sol is held in such a position that it can reflect light, an opalescence, like the bluishness of a soap solution, is often apparent. The appearance then is that the solution is not truly homogeneous and contains particles large enough to reflect or scatter the incident light.

This heterogeneity of colloid systems was realized to the full as a result of the invention of the ultramicroscope (*see* p. 76) by Siedentopf and Zsigmondy in 1903. In this instrument, no direct light passes through the objective, and the light scattered or reflected by particles present in the solution under examination is observed against a dark background. The result is that particles which are too small to be seen in the most powerful microscope with ordinary lighting arrangements are indicated by the halos of scattered light surrounding them. The limit of visibility of particles was thus extended from $500\mu\mu$ to $10\mu\mu$*. When colloidal solutions were examined by means of the ultramicroscope it became clear that they are not optically homogeneous like true solutions, but contain innumerable small

* μ is a convenient unit for measuring small particles and is equal to 0·001 mm. 1 $\mu\mu$ is $\mu/1000$, or 10^{-7} cm.

particles of the dispersed substance, the diameter of the particles being in general less than $100\mu\mu$. Such exceedingly small particles are very large, however, in comparison with the size of a molecule, and it immediately became apparent that the slow diffusibility and many other characteristics of colloids were consequences of the relatively large size of the particles of dissolved substance. On the other hand, the particles are too small to be retained by the hardest filter paper.

It gradually came to be recognized that colloidal solutions are heterogeneous systems consisting of two phases, one of which is in the form of extremely small particles, and such systems are therefore differentiated from true or molecular solutions, where the presence of two phases cannot be recognized. It should be borne in mind, however, that this distinction between homogeneous and heterogeneous systems is scarcely a real one. A solution of sodium chloride in water would be heterogeneous to an observer of molecular dimensions, who would regard water molecules, sodium chloride molecules, sodium ions, and chlorine ions as separate phases. Even water would be heterogeneous to an electron, and the Milky Way would appear homogeneous to an almost infinitely larger being. However, using the methods and instruments at our disposal, it is permissible to consider colloids as heterogeneous systems and ordinary molecular solutions as homogeneous systems, remembering that there is a continuous progression from one to the other and that we cannot say where heterogeneity begins. Thus, colloid chemistry becomes a province of *dispersoid* chemistry, or the chemistry of one phase dispersed in another.

The generalization of these ideas was made by Wolfgang Ostwald in 1907 (*Kolloid-Z.*, 1907, 1, 291, 331). He considered colloids as essentially two-phase systems, one phase being finely divided and dispersed in the other. The finely divided phase is in the form of discrete particles and is called the " disperse phase," and the other phase is a medium in which the particles are dispersed and is known as the " dispersion medium." Synonymous terms which are sometimes employed are " discontinuous and continuous phases " and " internal and external phases." In addition, the terms " micelle " and " intermicellar liquid " are sometimes used, though many authors attach a rather different significance to the word " micelle."

Ostwald showed further that colloids occupy an ill-defined intermediate position between coarse suspensions and true solutions, the difference being merely in the size of the particles of the disperse phase. He distinguished between (1) "coarsely disperse systems," which slowly settle under the influence of gravity and the particles of which are visible with the unaided eye or with an ordinary microscope; (2) "colloidally disperse systems," which do not settle under gravity and which are resolved only in the ultramicroscope; (3) "molecularly disperse systems," or true solutions, where the disperse phase consists of molecules or ions.

There is a perfectly gradual transition from one end of this scale to the other; from a pure salt solution to a suspension of sand in water. But just as a typical salt solution having a particle size of the order of $0 \cdot 1 \mu\mu$ has a characteristic series of properties, such as osmotic pressure and its effect on the freezing-point and boiling-point of the solvent, so when the particles of the disperse phase have diameters roughly between the limits $100 \mu\mu$ and $1 \mu\mu$ the system exhibits another series of properties which we have come to regard as characteristic of the colloidal state. It is convenient, therefore, to treat systems whose particle size falls roughly between these limits as a special study under the name "colloids," always bearing in mind that they pass gradually on the one hand into true solutions and on the other hand into coarse suspensions. Ordinary filter paper retains particles larger than about 5μ; the best microscope can detect particles of somewhat less than a tenth of this diameter, and the smallest particles observed under the ultramicroscope have been about $6\mu\mu$; the particles of a molecularly dispersed solution are about $0 \cdot 1\mu\mu$.

A nomenclature for the particles of dispersoids, depending on the degree of dispersion, has been proposed by Siedentopf and Zsigmondy. Particles visible in the ordinary microscope are called "microns," those which can be recognized in the ultramicroscope are termed "sub-microns," and those which are too small to be visible in the ultramicroscope are called "**amicrons**." Amicrons include the particles of very highly disperse colloids and also molecules and ions of ordinary solutes.

Degree of Dispersion. From these considerations it follows that the dispersity or degree of dispersion of the disperse phase is the essential feature of colloid systems. The finer the particles, the greater is the dispersity.

A very important study of the relation between the dispersity of a precipitate and the concentration of the reacting solutions was made by P. P. von Weimarn (*Grundzüge der Dispersoidchemie*, 1911), and has given experimental support to the theoretical considerations of Ostwald. One important generalization to be drawn from this work is that any substance sufficiently "insoluble" in the dispersion medium can be obtained in the colloidal state when suitable conditions prevail. Another is a simple relation between the concentration of the reacting salts and the fineness of division of the precipitate. In its simplest form, this can be expressed as

$$\delta = \frac{C}{S} \cdot \eta,$$

where δ is the dispersion coefficient representing the fineness of the particles, S is the solubility of the slightly soluble substance, C is the potential state of supersaturation, or the amount of substance that would be present in solution if it did not precipitate out, and η is the viscosity of the liquid.

The absolute values of C and S have no great significance, but the ratio C/S is a controlling factor and represents the force tending to drive the substance out of solution. C/S can be regarded as the momentary degree of supersaturation of the liquid with respect to the precipitate, and the equation shows that the greater the degree of supersaturation the higher will be the dispersion coefficient, or the smaller the particles. Consequently, with sufficiently high supersaturation particles of colloidal dimensions should be produced. A high viscosity of the liquid also favours a high degree of dispersion.

Von Weimarn's results on the production of precipitates of barium sulphate by mixing solutions of barium thiocyanate and manganese sulphate over a wide range of concentrations furnish a good illustration of this rule. Solutions of equal volume and equivalent concentrations were mixed, the concentrations being varied between the limits $N/20{,}000$ and $7N$. The results may for convenience be treated in four groups.

GROUP I. Range $N/20{,}000 - N/7{,}000$. $C/S = 0$ to 3 or slight supersaturation. The rate of crystallization was extremely slow, several years being required for the more dilute solutions, and large crystals were produced. This throws an interesting light on the large mineralogical specimens of barium sulphate.

GROUP II. Range $N/7{,}000 - N/600$. $C/S = 3$ to 48.

Complete precipitation took some months at the lower concentrations and a few hours at the higher concentrations. The product was a fine powder, which appeared crystalline under the microscope, the powder being finer the higher the concentration of the solutions.

GROUP III. Range $N/600 - 3N$. $C/S = 48$ to $88{,}000$.

Precipitation occurred in a few seconds at $C/S = 48$, but instantaneously at the higher concentrations. At

$$C/S = 20{,}000$$

the crystalline nature of the precipitate was scarcely any longer recognizable under the microscope. At higher concentrations the precipitate appeared amorphous to microscopical examination and in the highest region of the concentration range became gelatinous.

GROUP IV. Range $3N - 7N$. $C/S = 88{,}000$ to $200{,}000$.

These solutions formed viscous jellies of barium sulphate. The jellies were not permanent, the initial excessively small particles clustering in time into secondary aggregates.

These results demonstrate the validity of von Weimarn's formula and illustrate the effect of varying the term C in the equation. Similar results should also be obtained by varying the term S, or the solubility of the substance. With barium sulphate this can be done by the addition of alcohol, which reduces the solubility of barium sulphate considerably below its already small value. Von Weimarn found that the addition of alcohol to the solutions does increase the dispersity of the precipitate and that more stable sols of barium sulphate can be prepared in this way.

It is possible, then, to formulate conditions which will determine whether the system produced will be molecularly disperse, colloidally disperse, or coarsely disperse.

Types of Disperse Systems. Having arrived at the generalized concept of dispersoids as polyphase systems, Ostwald

AS POLYPHASE SYSTEMS

went on to consider the possible types of such systems that could exist, and adopted a classification based on the states of aggregation (solid, liquid, or gas) of the phases. So far, in our brief references to colloid systems, we have considered only diphasic systems having a solid as a disperse phase in a liquid (water, in fact) as dispersion medium, but the generalized concept of dispersoids includes all possible polyphase systems, where the disperse phase is present in the form of small enough particles. The complete list of the possible colloidally disperse systems is as follows, the dispersion medium being placed first and the disperse phase second.

(1) Gas-liquid . . . . . . Mist.
(2) Gas-solid . . . . . . Smoke.
(3) Liquid-gas . . . . . . Foam.
(4) Liquid-liquid . . . . . . Emulsion.
(5) Liquid-solid . . . . . . Suspension.
(6) Solid-gas . . . . . . Solid foam.
(7) Solid-liquid . . . . . . Solid emulsion.
(8) Solid-solid . . . . . . Solid sol.

Colloid systems of two gases cannot exist because two phases cannot be produced, gases being completely miscible. A gas-liquid system can be described as a mist and is exemplified by clouds and sprays. A gas-solid system can in general be termed a smoke and is often produced by the rapid condensation of a vapour. For example many metals or metallic oxides condense from the vapour to form smokes. Coarsely disperse systems of this type are also formed by the direct dispersion of a finely-divided solid in air. A good example is the solid fog produced when flour or other finely divided material is shaken up in the air. A liquid-gas system is a foam, and the classical example is the froth on beer. Liquid-liquid systems are called emulsions and these are very important. They are typified by the white emulsions formed by shaking oils with water, generally with the addition of a small quantity of other substance to keep the emulsion stable. The liquid-solid system of colloidal dimensions is called a suspensoid sol and to this class belong colloidal gold and the greater number of colloids with which we have to deal. A solid-gas system may be called a solid foam—a type encountered in pumice. A solid-liquid system or solid emulsion is not common, but is encountered in some minerals which contain finely dispersed liquids. On the other hand, jellies may belong to this type. Colloid systems of two solid phases are found in some alloys, many coloured precious stones, stained glasses, and blue rock salt.

Although the study of colloids embraces the widest number of physically heterogeneous systems, experimental and theoretical work has so far been confined mainly to two of these types—the liquid-liquid and liquid-solid systems, or the emulsoid and suspensoid sols as they are called. The type to which the jellies formed from these sols belong is not yet quite clear, as the ultimate structure of the jellies is not completely understood.

When the particles of the disperse phase are exceedingly small it is often difficult to decide whether they are to be regarded as having liquid or solid properties. The total surface of the disperse phase is then so great that surface tension becomes an overwhelming factor and may give to excessively small droplets of liquid something of the rigidity of a solid, or to minute particles of solid the surface properties of a liquid. Because of facts such as these, Zsigmondy prefers to be content in the first place with a consideration of three types of colloid systems, according to whether the dispersion medium (the state of aggregation of which is at once apparent) is a gas, a liquid, or a solid.

However, for the purpose of this book it will be sufficient to confine detailed attention to systems where the dispersion medium is liquid and the disperse phase is either liquid or solid, although in some cases the state of the disperse phase is not known with accuracy. These two types have been studied in much greater detail than any of the others by a large number of investigators and the generalizations of colloid chemistry refer to them. It would not be correct to apply these generalizations to all the remaining types of disperse systems, though in certain cases this may be done.

The greater part of our study of the two types will be restricted to systems where the dispersion medium is water, because of the pre-eminent importance of these systems. Colloidal solutions in water are termed " hydrosols." Some investigators have made studies of colloidal solutions where the dispersion medium is alcohol (" alcosols "), or benzene (" benzosols ") or some other liquid. Sols in organic liquids as the dispersion medium are generally called " organosols."

From the practical point of view water is the most important dispersion medium, but it can also play the part of a disperse phase. It does so in clouds and mists, where air is the dispersion medium. In the highest fleecy clouds the particles are believed to be in the form of ice, giving a gas-solid system with water as the disperse phase. It is also very easy to prepare emulsions of water dispersed in oil, and by freezing solutions of water in alcohol the ice may be made to form in the state of colloid particles. N. von Weimarn has recently obtained colloidal ice by chilling concentrated sugar solutions at $-80°$ C. The velvety feel of ice-cream is said to be due to the ice separating out in the form of particles of colloidal dimensions through the addition of a colloid such as gelatin to the mixture. If the particles are considerably larger the tongue can detect the roughness.

CHAPTER III
THE FORMATION OF COLLOID PARTICLES

Many substances dissolve spontaneously in water, either in the cold or on warming, to give colloidal solutions directly. Gelatin, agar, starch, gums, soaps, etc., are examples of this class. Other substances, whilst not dispersing directly in pure water, may do so in salt solutions, particularly in solutions of very soluble salts such as calcium thiocyanate. Silk and cellulose belong to this class; cotton dissolves to form a sol in aqueous cuprammonium hydroxide. These substances are sometimes regarded as " natural " colloids, and those systems, such as colloidal gold, which have to be prepared by special means are often referred to as " artificial " colloids, though such a division is not strictly logical. The so-called natural colloids are in general substances of high molecular weight, and in some instances may form molecular solutions having colloidal properties simply because the molecule itself is of colloidal dimensions. The distinction between molecularly disperse and colloidally disperse systems becomes somewhat vague in such cases.

The so-called artificial colloid systems of substances of relatively low molecular weight are produced by special means and the general methods available have been studied particularly by Svedberg (*Formation of Colloids*, London, 1921). Since colloid particles occupy an intermediate position between true solutions and coarse suspensions of massive material, there are two processes by which they may be obtained; either by increasing the degree of dispersion of massive particles or by decreasing the degree of dispersion of the molecularly dissolved particles. Svedberg calls these processes " dispersion " and " condensation " respectively. So a colloid system can be produced either by breaking up massive material into particles of the requisite size or by causing molecules to unite to form large enough particles.

THE FORMATION OF COLLOID PARTICLES

Dispersion Methods

Mechanical Dispersion. Mechanical methods of dispersion in gases generally give rather coarse-grained systems and, as in all cases of dispersion, the particle size is by no means uniform. Regener has adapted a method of disintegrating a liquid by means of a gas-jet so as to give a disperse liquid phase of fairly uniform particle size. These sprays are probably produced by the disruption of thin lamellæ of the liquid formed between two gas spaces. In the laboratory, Regener's spraying apparatus is often useful in producing coloured flames for spectroscopic work. The spray is made from a salt solution and is led into a colourless Bunsen flame. In industry the spraying process is applied to the purification of air or other gases from dust particles; the fine spray gradually settles, bringing down with it the dust particles, and the process is invaluable in certain catalytic reactions, where the catalyst might become poisoned, and in such industrial operations as the manufacture of films for photographic purposes, where the air must be perfectly free from dust.

Disperse systems of solids in gases may be produced by forming a gas under high pressure in a solid, which disrupts explosively when the pressure is reduced suddenly. Volcanic dust is formed in this way. The more usual way of producing this type of system is by the sudden chilling of vapours, such as occurs in many metallic arcs, but the mechanism of that process is strictly one of condensation.

The disperse systems in gases have not been considered very fully from the standpoint of colloidal behaviour, but the process of dispersion in liquids has already received a considerable amount of study.

The most fundamental method has been developed in quite recent years and consists of grinding the material to be dispersed in what is known as a " colloid mill." The principle of this machine is as follows. The substance to be dispersed is suspended in the form of coarse particles in the dispersion medium and passes through a channel in which are two metallic discs placed close together and rotating at high speed in opposite directions. During passage between the two discs, the particles of the suspension are torn so as to yield particles of colloidal dimensions. For this purpose it is essential to have a very high speed

—about 30 m. per sec. Another feature is that it is generally necessary to add some other substance to the initial suspension in order to prevent immediate coagulation of the sol formed.

Other less general methods of mechanical dispersion generally require some special chemical treatment in order to keep the disperse system stable. Kužel succeeded in preparing sols of various metals by grinding them very finely and then treating the product alternately with acid, alkali, and pure water. A point of particular interest about these sols is that he succeeded in coagulating them and used the coagulum for making filaments for electric lamps. This process has no commercial importance at the present time.

Liquid-liquid systems, such as oil and water, are relatively easy to disperse by simple mechanical agitation, but as a rule the resulting emulsion is not stable unless a small quantity of a third substance known as an emulsifier (sometimes called dispergator) is also present. The emulsifier may be an electrolyte or a colloid; soap, which is both, enables many oil-water systems to emulsify easily on vigorous shaking. One important influence of the emulsifying agent is certainly to lower the interfacial tension between the two liquids, so favouring the increase of surface which accompanies dispersion, but another is probably an electrical effect which will be better understood later on.

Peptization. When solutions of ammonium carbonate and ferric chloride are mixed, ferric hydroxide is precipitated. If the freshly precipitated ferric hydroxide is treated with a small quantity of ferric chloride solution it disperses immediately to form a dark reddish-brown colloidal solution. We call this process peptization and the ferric chloride the peptizer or peptizing agent. If the last traces of peptizing agent are removed by exhaustive dialysis, the sol becomes unstable and eventually coagulates.

There are numerous other examples of the peptization of precipitates to give sols. Freshly precipitated stannic oxide (prepared by mixing solutions of sodium stannate and sodium hydrogen carbonate) washed and suspended in water forms a clear sol of stannic acid on addition of a few c.c. of ammonia. Many freshly precipitated metallic sulphides, washed and suspended in water, can be peptized to clear sols by bubbling hydrogen sulphide through the suspension.

The amount of peptizing agent required depends on the treatment and method of preparation of the precipitate, and since freshly prepared precipitates peptize most readily it seems very probable that precipitates which are capable of peptization already consist of particles small enough to form sols, which have been loosely aggregated in the process of precipitation. The peptizing agent separates these ultimate particles, probably by giving them an electric charge. As will be apparent later, electric charges on colloid particles are of fundamental importance in determining the stability of the system. However, if this view of peptization is correct, the process is not one of dispersion in the sense of dividing up structural units of matter, but rather a separation of the units; therefore some authors prefer to regard the process as reversed coagulation and not dispersion. Nevertheless, since we start with a massive precipitate and end with a sol, it is convenient to regard it as dispersion.

Peptization is often brought about by adding a small quantity of electrolyte, which may act in the way suggested above, but sometimes precipitates are peptized by thorough washing with distilled water. When precipitates are produced by double decomposition there is generally produced simultaneously a soluble reaction product. When exceedingly dilute solutions are used, the insoluble product may form a sol, but generally such sols are precipitated or coagulated by the soluble electrolyte, which as a rule adheres to the precipitate. Such an electrolyte is said to be *adsorbed*. Thorough washing with distilled water may remove a considerable portion of the adsorbed electrolyte and so enable the particles to disperse once more to form a sol. In general, a trace of electrolyte is necessary to keep a sol stable, but larger quantities of electrolytes coagulate the sol. Peptization by removal of a coagulating electrolyte is the cause of the troublesome phenomenon often encountered in analysis, that the precipitate goes through the filter paper on thorough washing.

Electro-dispersion. In 1898, Bredig (*Z. angew. Chem.*, 1898, 951) described an electrical disintegration method for the preparation of metal hydrosols. The method consists of forming an electric arc between two wires of the metal immersed in "conductivity" water. The general arrangement of the apparatus is shown in Fig. 1. The current from a direct current lighting main is led through a suitable resistance *R* and an ammeter *A* to the gold wires *a* and *b*. For convenience in handling, these are sheathed in glass tubing. An arc is struck between the gold wires by touching them momentarily whilst under water and drawing them apart a short distance. To maintain the arc under water it is generally necessary to repeat this process fairly often. The heating effect of the arc is inimical to the stability of the sol and the preparation vessel should therefore be cooled in ice. In many cases it is advantageous to add a trace of alkali to the water, but other electrolytes should be absent.

Using gold electrodes, this treatment gives rise to a purplish sol. The dispersed metal comes from the cathode, and the

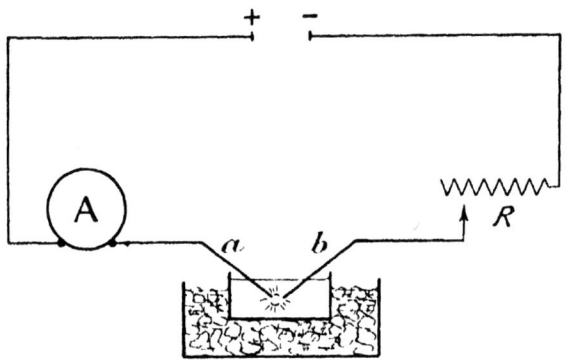

FIG. 1.—Electro-dispersion of Metals.

anode generally gains slightly in weight. Platinum gives a black or brown sol, silver a brown or brownish-green sol, mercury a grey or black sol, whilst more easily oxidizable metals such as zinc and copper give rise to sols whose particles consist mainly of oxide. The method is most generally useful for the preparation of hydrosols of the noble metals; it is not likely that a metal which is less noble than hydrogen will form a hydrosol, but rather that the particles will consist of the oxide.

When direct current electro-dispersion was applied to the production of sols of metals in organic dispersion media such as alcohol, difficulty was encountered through the decomposition of the organic liquid. In some cases the dispersed product was found to contain as much as 70 per cent. of carbon. This difficulty has been overcome and the general application of the method has been developed more extensively by Svedberg, who has made numerous improvements on Bredig's original method (*Colloid Chemistry*, New York, 1924).

One improvement is to form the arc between two metal wires enclosed in a silica tube which has a hole bored just in front of the arc. The silver vapour produced in the arc is blown out through the hole by a stream of nitrogen and condenses in the liquid medium to give a sol. In this way a higher degree of dispersion is obtained and the liquid dispersion medium is not decomposed by the arc. The diameter of the particles produced by the free arc is generally about $40\mu\mu$, whilst that of the particles produced by the enclosed arc is about $5\mu\mu$. This gives rise to a difference in the colour of the sols. An unfortunate feature is that the cooling is not so effective as in the free arc, and the silica tube gets so hot that it gives off its own vapour and colloidal silica appears in the product. For the same reason, the enclosed arc cannot be used for preparing sols of those metals which have a very low melting-point, for the electrodes fuse together too readily.

Svedberg has also made a close examination of the electrical dispersion of metals by means of an alternating current arc and has studied the particle size and distribution in its relation to the frequency of the alternation and the influence of the kind of spark gap. The high frequency oscillatory arc obtained from an induction coil and suitable condenser was found to give a more highly dispersed and a purer product than that obtained from the direct current or low frequency alternating current arc. Low self-induction, high capacity, and short length of arc favour the formation of pure sols.

It has been possible to prepare colloidal solutions of practically all the metals by using the oscillatory arc. Ether may be used as dispersion medium for the more reactive metals and in this liquid sols of the alkali metals have been produced. In this method, the electrodes are generally platinum wires sealed into the glass vessel and the metal to be dispersed is placed loosely between them in the form of small pieces. If the metal used is much more easily dispersed than platinum, the sparking between the platinum wire and the nearest piece of the other metal can be neglected. Otherwise, electrodes of the same metal have to be used.

Using carefully dried ether as the dispersion medium and ensuring the absence of oxygen by passing a continuous stream of hydrogen, Svedberg has developed a method for obtaining sols of the alkali metals. The ether is strongly cooled in order to lower the velocity of whatever chemical reactions might take place. The pieces of the alkali metal are placed in a loose pile between two platinum wires. The sodium sol prepared in this way is purple in transmitted light.

Svedberg regards this general method of electro-dispersion as a condensation process, and it is true that the actual formation of the sol particles is one of condensation from the vapour of the metal which is first produced in the arc. However, dispersion to a vapour precedes condensation to colloidal particles, and since the total effect consists of starting with massive material and ending with a colloidal product it is reasonable to regard the whole process as one of dispersion. Actually, some direct dispersion to particles of colloidal size may occur simultaneously, for cathodes of soft metals like tin and lead disperse in alkali solutions to form hydrosols under ordinary conditions of electrolysis if a high current density is employed. A tellurium cathode disperses to a sol directly in pure water when under the influence of a fairly high potential.

Dissolution of Salts. A transient stage of colloidal dispersion in the ordinary process of dissolution of salts in water has been described by Traube (*Kolloid-Z.*, 1929, **47**, 45). When a drop of saturated potassium dichromate solution is allowed to crystallize on a microscope slide and a drop of water is then placed in contact with it, ultramicroscopical examination reveals the presence of numerous submicrons. They appear to be produced by the disintegration of the crystals of potassium dichromate prior to their dissolution. The same phenomenon is observed with most other salts which are not too soluble, particularly with mercuric chloride, where the submicrons may have a life-period of one or two minutes. Complex compounds, such as potassium hexanitrocobaltate, show the effect best. The phenomenon is considered to be consistent with Smekal's theory of "lattice blocks," according to which crystals are built up of blocks of molecular units separated by pores or canals. The lattice blocks and the submicrons have the same order of magnitude, about 10^{-5} mm.

CONDENSATION METHODS

These include all processes by which colloidal systems are produced from systems in which the substance is originally present in a molecularly or ionically dispersed form.

Reduction. Sols of the noble metals can be made by reducing very dilute solutions of their salts, made either neutral or slightly alkaline, with a large variety of reducing agents. The favourite agents are hydrazine, phenylhydrazine, pyrogallol, hydroquinone, formaldehyde, hydrogen peroxide, etc., but investigations have included a very wide assortment of substances such as alcohols, sugars, saliva, urine, macerated leaves and petals, tobacco juice, and other vegetable extracts. A great deal of interest, historical, experimental, and theoretical, is attached to the formation of gold sols by reduction of chloroauric acid.

Colloidal gold has been known for more than two centuries in the form of ruby glass and purple of Cassius, though it was not recognized until comparatively recent times that the gold is in a colloidal form in these preparations. Ruby glass is made by adding a gold compound to a suitable glass and heating, when an almost colourless glass results; but if this be cooled and reheated, ruby glass is obtained. Probably the first stage consists of a reduction of the gold compound to a very highly supersaturated, colourless, molecular solution of gold in the glass, and the second stage is a recrystallization of the gold on annealing to form particles of colloidal dimensions.

⁺ The production of hydrosols of gold by reduction of chloroauric acid has received extensive study, particularly by Zsigmondy (*Chemistry of Colloids*, Chapman & Hall) and by von Weimarn (*Rep. Imp. Ind. Inst. Osaka*, 1929, 9, No. 7, 9). Early work on gold hydrosols was done by Faraday (*see* p. 3), but this did not attract much attention until about the beginning of the present century, when Zsigmondy (*Annalen*, 1898, **301**, 29) rediscovered what Faraday had already found and carried his investigations much further.

In general, the colloidal reduction product is purple or blue, but when special precautions are used pink or ruby coloured gold sols may be obtained. The particles of the red sols are generally smaller than those in the blue sols and in some cases the dispersion of the gold may be so high that the presence of particles cannot be recognized in the ultramicroscope. The gold is precipitated out by quite small quantities of electrolytes and the presence of the minutest trace of electrolyte is sufficient to cause a change in the state of aggregation of the particles of red gold sols, so that they turn blue or violet. For this reason, not only is it essential to use water and other materials of the highest possible degree of purity, but all reaction vessels must be made of hard glass and should be subjected to thorough steaming before use.

Von Weimarn (*Kolloid-Z.*, 1928, **45**, 366) has given directions for the production of orange gold sols. Solutions of auric chloride and sodium citrate are mixed together in boiling distilled water and a small quantity of dilute potassium cyanide solution is subsequently added. The whole is then kept boiling for a period of many hours. Excess of potassium cyanide gives a quite colourless solution. The sols remain stable for some months; they generally become pure red in time. The change from orange-red to orange on the addition of potassium cyanide is considered to be caused by the greater velocity of dissolution of the red colouring particles of gold. Orange gold sols are also described, the colour of which is due to the suspension of a red sol in a yellow dispersion medium (often produced by the action of alkali on reducing substances such as sugar, etc.), and it is pointed out that the colour of a sol is not a safe guide to its degree of dispersion.

Zsigmondy gives the following directions for the preparation of stable hydrosols of gold. 120 c.c. of specially distilled water are brought to boiling in a Jena glass beaker and meanwhile 2·5 c.c. of a 0·6 per cent. solution of $HAuCl_4, 4H_2O$ and 3·5 c.c. of 0·18 N- potassium carbonate are added. When the liquid boils, 3 to 5 c.c. of 0·12 per cent. formaldehyde are added, and the liquid stirred vigorously. Soft stirring rods must be avoided. In a few seconds an intense red colour appears and does not change on further standing. The specially pure water is obtained by redistilling water in a silver apparatus and collecting the water from the silver condenser in a Jena glass receptacle. Svedberg does not recommend the use of silver stills for preparing pure water, for he finds that water which has been in contact with silver always contains some kind of silver compound, which is reduced by light to colloidal silver. The particles of these gold sols are generally visible in the ultramicroscope and their size is usually from 10 to 40 $\mu\mu$. Amicroscopic particles may be obtained by using more dilute solutions.

Another method of obtaining gold sols having amicroscopic particles is a development of Faraday's original method. Using the solutions of chloroauric acid and potassium carbonate recommended above, the mixture is reduced by adding about 0·5 c.c. of a solution of phosphorus in ether. This solution is best prepared by saturating ether with phosphorus and then diluting with ether to five times the volume. The mixture is allowed to stand for 24 hours, in which time it has gradually become red. The process may be hastened by boiling the liquid.

By a combination of these two processes it has been possible to procure gold sols containing particles of almost any desired size and further reference will be made to this later (p. 34).

Less highly disperse gold sols were obtained by Gutbier by reducing chloroauric acid with hydrazine, phenylhydrazine, or hydroxylamine. A deep blue sol can be obtained by adding to a litre of 0·1 per cent. chloroauric acid, neutralized with sodium carbonate, a few c.c. of a 0·025 per cent. solution of hydrazine hydrate.

One of the simplest reductions is that by hydrogen peroxide, which follows the equation

$$2HAuCl_4 + 3H_2O_2 = 2Au + 8HCl + 3O_2.$$

This reaction can be followed by measuring the increase in electro-conductivity due to the free hydrochloric acid.

After colloidal gold, silver hydrosols have received most attention as examples of the reduction process. They may be obtained by the action of many reducing agents on very dilute ammoniacal solutions of silver oxide. Their colour may be blue, green, yellow, or brown, according to the method. A very simple reduction method for preparing colloidal silver was worked out by Kohlschütter (*Kolloid-Z.*, 1913, **12**, 285) and consists of reducing a saturated aqueous solution of silver oxide by means of hydrogen. Excess of solid silver oxide is added and the temperature is kept between 50 and 60°. By passing hydrogen through a filtered saturated aqueous solution of silver oxide at 60° a yellow silver sol can be obtained in about 30 minutes. The sols contain some unaltered silver oxide, but if they are transferred to a platinum vessel and hydrogen is passed again the dissolved silver oxide is reduced to the form of silver crystals which adhere to the walls of the vessel and a pure silver sol is left.

Carey Lea's silver sols (*Amer. J. Science*, 1889, **37**, 476) have attracted considerable attention. He reduced silver nitrate with mixtures of ferrous sulphate and sodium citrate. The silver sols produced are far from pure as they contain organic substances, but they are remarkable in at least two respects. The majority of sols of metals can be obtained only in low concentrations—seldom as much as 0·1 per cent., but 10 per cent. silver sols can be obtained by Carey Lea's method and they are very stable. Again, it is characteristic of metal sols that when they are coagulated or precipitated the sol cannot be reproduced from the precipitate; that is to say, the coagulation of these sols is generally an irreversible process. When, however, the Carey Lea's sol is precipitated by ammonium nitrate it redissolves to a sol in water and precipitation with alcohol gives a product which contains about 99 per cent. of silver. This product, even after drying, redisperses in water to form a sol; the coagulation of these sols is therefore reversible.

Most other easily reducible metals, platinum, palladium, mercury, etc., can be prepared in colloidal form by these reduction processes suitably modified for the special case. They are not as a rule so stable as colloidal gold. Many of the sols are coagulated by boiling and they are all precipitated by small amounts of electrolytes.

Oxidation. One or two oxidation reactions which furnish the product in colloidal form are of interest; for example, selenium sols are produced by the oxidation of solutions of hydrogen selenide and sulphur sols result from the oxidation of hydrogen sulphide solutions either by oxygen or by sulphur dioxide. The simplest representation of the last reaction is

$$2H_2S + SO_2 = 3S + 2H_2O,$$

but this does not represent all that happens, for it is well known that pentathionic acid and other sulphur acids are produced simultaneously. Odén (*Nova Acta, Upsala*, 1913, **3**, No. 4) has made a careful study of this reaction from the colloid chemical point of view and has found that the amount of colloidal sulphur produced and its degree of dispersion depend on the concentra-

COLLOID PARTICLES

tion of the sulphur dioxide solution into which the hydrogen sulphide gas was passed. It was found that the amount of colloidal sulphur formed is greater at the higher concentrations and that the degree of dispersion also increases with the concentration. Thus, the formation of colloidal sulphur is favoured by conditions under which pentathionic acid is produced, and at low concentrations, where the reaction takes the course represented by the equation given above, the sulphur is precipitated. The presence of pentathionic acid therefore seems to be necessary for the stability of these sols.

Dissociation. Dissociation methods are not very important for the preparation of colloids, but nevertheless have considerable practical interest since it is believed that this process is a fundamental one in photography. Some study has been made of the dissociation of solid salts, particularly halides. Lorenz found that crystals of the chlorides of lead, thallium, and silver prepared in the ordinary way always contain ultramicroscopic particles which appear to be free metal. The substance can be freed from such particles only by passing through it in the molten state a mixture of dry chlorine and hydrogen chloride. When these "optically empty" crystals were exposed to light in the ultramicroscope, particles appeared, and there seems to be no doubt that they are formed by a dissociation process. Systems of the same kind can be produced by adding a small piece of metallic lead to molten lead chloride. When crystals of silver bromide are exposed to illumination, ultramicroscopic particles of metallic silver are produced. If the silver halides are exposed to feeble illumination no such particles appear in the ultramicroscope, but reduction to the metal occurs when the halide is treated with a suitable reducing agent or developer. In this case it is probable that amicroscopic particles are produced by the action of light and that development starts from these nuclei. The production of solid sols by a dissociation process is therefore an important matter in photography.

Another interesting example is the formation of blue rock salt. Occasionally, blue rock salt is found in nature, but it can be produced artificially by bombarding pieces of colourless sodium chloride with cathode rays. The blue substance shows the presence of ultra-microscopic particles, and these are probably metallic sodium, since the same kind of blue rock salt can be produced by heating crystals of colourless rock salt in a vacuum and then exposing to sodium vapour.

The application of dissociation to the preparation of a colloid is instanced by the production of a nickel benzosol. When nickel tetracarbonyl is heated it decomposes with separation of metallic nickel according to the equation

$$Ni(CO)_4 = Ni + 4CO.$$

If a solution of the compound in benzene is heated it dissociates in the same way, the carbon monoxide escapes and a benzosol of nickel is obtained. This has a black or brown colour.

Other examples are perhaps the formation of hydrosols of sulphur by adding dilute acid to dilute solutions of sodium thiosulphate or of polysulphides. In the first instance it is generally supposed that thiosulphuric acid is formed momentarily and dissociates into sulphur, sulphur dioxide, and water. In the second case it is often held that the first product is a hydrogen polysulphide, such as H_2S_2, which dissociates into hydrogen sulphide and sulphur.

Exchange of Solvent. When a solution of a substance is added to a large volume of a liquid with which the solvent is miscible, but in which the solute is practically insoluble, the solute is precipitated out in a very highly disperse condition. For example, gum mastic and gamboge dissolve in alcohol to give a clear solution, but they are not soluble in water. By adding a small quantity of the alcoholic solution to a large volume of water and stirring an opalescent hydrosol is obtained. Sols of sulphur or phosphorus are readily obtained by adding saturated solutions of these elements in alcohol to a large quantity of water. By choosing suitable dispersion media and solutes this method is capable of very wide application.

A variation is the addition of a soluble complex salt to a large volume of water in which one of the components is insoluble. Silver iodide is insoluble in water, but is soluble in a concentrated solution of potassium iodide. When the solution in potassium iodide is poured into a large volume of water the silver iodide is rendered insoluble and is obtained as a sol. A similar effect is often obtained in the qualitative analysis of a substance containing silver. If for any reason the substance has had to be dissolved in concentrated hydrochloric acid some of the silver chloride

COLLOID PARTICLES

dissolves and on dilution before proceeding to Group II a milkiness appears and generally escapes filtration. This is due to colloidal silver chloride.

Double Decomposition. In accordance with von Weimarn's reasoning, it should be possible to prepare sols of many substances by chemical precipitation where the solubility of the substance is sufficiently small and the potential supersaturation high. Actually, there are more factors to consider than these in such reactions, because in a double decomposition a soluble reaction electrolyte is generally produced in addition to the insoluble substance. This soluble electrolyte plays a very important rôle, because one general feature of the sols under discussion is that they are coagulated or precipitated by quite small concentrations of electrolytes. Consequently, even if the conditions were such as to produce a highly dispersed precipitate of colloidal dimensions, that system might only exist momentarily, being soon coagulated into coarse secondary particles by the electrolyte present.

The argument leads to two conditions under which the formation of sols is to be expected in double decompositions: (*a*) where no reaction electrolyte is produced, and (*b*) where the reacting solutions are so dilute that the reaction electrolyte does not reach the concentration required for coagulation to begin. Against the second condition is the fact that with increasing dilution the chance of formation of an initially highly disperse system diminishes unless the substance is extremely insoluble, leaving as the main condition the formation of no reaction electrolyte or of an electrolyte which is feebly dissociated.

A good example is the production of arsenious sulphide sol by the action of hydrogen sulphide on arsenious acid. Here no reaction electrolyte is formed. To prepare this sol, 1 grm. of arsenious oxide is dissolved in 500 c.c. of water by keeping the latter boiling until dissolution is complete. (This may take some time and means occasionally replacing the evaporated water by fresh distilled water.) The liquid is cooled and then a slow stream of hydrogen sulphide is passed through until the colour no longer deepens. The excess of hydrogen sulphide is removed by passing a stream of hydrogen through the sol. The sol has a deep yellow colour.

An example of the production of a sol simultaneously with a

weakly dissociated electrolyte by double decomposition is the formation of mercuric sulphide sol by passing hydrogen sulphide into a dilute solution of mercuric cyanide. In this case, all the solutes present, hydrogen sulphide, hydrogen cyanide, and mercuric cyanide, are feebly ionized and the sol remains stable because it is not under the influence of coagulating agents.

In many reactions of double decomposition sols are produced and remain for varying lengths of time, but more or less rapidly coarsen through electrolyte coagulation.

Hydrolysis. So far as hydrosols are concerned, hydrolysis is a double decomposition involving the dispersion medium as a reactant. Naturally, only oxides, hydroxides, and basic salts can be produced in colloidal form by this means, but the method is general so far as it relates to an important group of sols.

Ferric hydroxide sol may be prepared by adding a few c.c. of 30 per cent. ferric chloride solution to 500 c.c. of boiling water. The liquid turns deep red and remains so when cooled. It contains colloidal ferric hydroxide and free hydrochloric acid produced by hydrolysis. The hydrochloric acid may be removed by dialysis, but not completely, for if the last traces are removed the sol becomes unstable and coagulates. A trace of electrolyte is necessary for its stability. The dark red colour of ordinary solutions of ferric chloride does not correspond to the colour of ferric ions and is due to partial hydrolysis giving colloidal ferric hydroxide. This sol is a favourite one for colloid-chemical investigations.

The Process of Condensation

In all these condensation processes the substance which is finally obtained in particles of colloidal size is either initially in a molecularly dispersed state or passes through that state under a very high degree of supersaturation. For instance, when molecules of gold chloride are being reduced so as eventually to form colloidal gold, at a certain moment there is present a highly supersaturated solution of gold molecules, which later condense to form the colloid particles. At least, it is reasonable to suppose that this is what happens.

Viewed in this light, the process of formation of the colloid particle by condensation is one of crystallization from a supersaturated solution and the dispersity of the product will be determined by the laws governing the formation of crystallization centres. Whether the substance will remain stable at its initially high dispersity and continue as a sol, or whether shortly after its formation a further change in dispersity in the direction of coarseness will occur, is another important factor which depends largely on electrical conditions such as the presence of ions.

Leaving the electrical considerations out of mind for the present the formation of a new solid phase from a liquid phase has received considerable investigation, particularly at the hands of Tammann and his pupils (*Kristallisieren und Schmelzen*, Leipzig, 1903). The fundamental facts are that the process must be differentiated into two parts : the formation of crystallization nuclei and the velocity of crystallization. The first part of the process is the union of groups of molecules to form centres of crystallization, and the second part is the growth of these nuclei at the expense of the material still in molecular dispersion.

It has been possible to count the number of crystallization nuclei produced by various degrees of supercooling of melts and the results of these investigations show that the number of nuclei increases at first as the temperature is reduced below the melting point of the substance, passes through a maximum and finally decreases. The form of the curve suggests that the phenomenon is governed by probability and it seems likely that at any place where a large number of atoms are close together in a configuration resembling a space lattice, a crystal nucleus is formed. The formation of nuclei is also affected, sometimes in one direction and sometimes in the other, by the presence of foreign bodies, whether heterogeneous or homogeneous.

The rate of crystallization of the nuclei, once they have formed, increases at first as the temperature is reduced below the melting point, remains constant over a certain range of temperature, and finally falls rapidly.

The point of all this is that if conditions are such that the rate of formation of nuclei is small and the rate of crystallization large, then a few large crystals will be produced and highly disperse systems are quite impossible. If, on the other hand, there is very rapid formation of nuclei and a low rate of crystallization a large number of small particles will be produced and in favourable cases stable sols may be formed.

From what has been said about peptization it is already apparent that there are other conditions affecting the ultimate dispersity of a precipitate, but these are secondary to the two conditions which govern the actual condensation process.

CHAPTER IV

GENERAL CHARACTERISTICS OF COLLOID SYSTEMS

Osmotic Pressure and Molecular Weight. The first criterion of colloidal nature was the inability of colloids to pass through a membrane which allowed the diffusion of ordinary salts. Quite apart from any other evidence, this might be taken to mean that the particles of colloids are larger than those of crystalloids and consequently cannot make their way through the membrane. This idea immediately receives support from the fact that those substances which are known for chemical reasons to have a particularly high molecular weight (gelatin, starch, gums, etc.) generally give colloidal solutions spontaneously. The molecular weight of hæmoglobin, the red colouring matter of blood, is deduced from chemical evidence to be roughly 16,300. Substances of such high molecular complexity will tend to form sols with colloidal properties even if they are molecularly dispersed, for the molecule itself is a colloid particle. It must be remembered also that there is no reason why under suitable conditions these large molecules should not fit into a space lattice, and indeed it is possible to prepare crystals of many of these substances which are always regarded as natural colloids. This reminds us that we must think of colloid systems rather than colloid substances.

Turning to the simpler hydrosols of metals, metallic sulphides and hydroxides, etc., we have to distinguish between the chemical molecular weight and the physical molecular weight. The chemical molecular weight is that indicated by the chemical formula of the molecule and this is naturally comparatively very low in most of the colloids belonging to the artificial type. The physical molecular weight is indicated by osmotic pressure and allied phenomena. As we have already seen, it must be supposed that the particles of sols of metals and similar colloids consist of large aggregates of molecules.

From the kinetic point of view, these large aggregates of molecules do not behave differently from molecules or ions; they exert an osmotic pressure from which, if it could be measured, their " molecular weight " or aggregate weight could be calculated. The osmotic pressure of all systems can be calculated from the formula

$$P = \frac{RT}{N} \cdot n$$

where n is the number of particles in unit volume and N is the Avogadro constant or the number of particles in 1 gram-molecule, i.e. $6 \cdot 06 \times 10^{23}$. The essential feature of this formula is that the osmotic pressure is determined not by the size of the particles, but by the number of particles, whether ions, molecules, or coarse aggregates, per unit volume. Thus, with increasing dispersion of a given weight of substance the osmotic pressure increases, because n becomes larger.

However, when applied to colloidal solutions, osmotic pressure does not seem to be very helpful in practice. We may calculate what would be the osmotic pressure of a gold sol with particles having a mean diameter of $4\mu\mu$. This is a very favourable case, for these particles are amicrons. Suppose the concentration of the sol is 0·1 per cent. Then if each particle measures $4\mu\mu$ across, there are 10^{18} such particles in a litre of the sol. A sol containing $6 \cdot 06 \times 10^{23}$ particles per litre can be called a " normal solution " from the point of view of the aggregates, and the sol under discussion is therefore 0·000002 normal, and its osmotic pressure is about 0·5 mm. of water and corresponds to a lowering of the freezing point by 0·000004°.

It is obvious that such small changes as these are quite useless in determining the molecular weight of the colloid. Colloids can therefore be expected to have practically no measurable osmotic pressure and no effect on the freezing-point, boiling-point, or vapour pressure of the liquid dispersion medium.

When experiments are carried out on these lines, the conclusions are borne out in that any effects observed are small compared with those obtained with true solutions, but many colloids exert a much greater osmotic pressure than would be expected. Whether this is a real property of the colloid is not certain. It must be remembered that colloids practically always contain a peptizing electrolyte and a trace of free electrolyte will swamp whatever effect is due to a good deal of colloid. The electrolyte adsorbed on the particles is of course in equilibrium with some free electrolyte in the dispersion medium. It may be argued, however, that since the electrolyte is an essential constituent of the colloid system we must regard the observed osmotic pressure as a true value expressing that particular system, consisting of the colloid particles plus the stabilizing electrolyte. In some cases, fairly reproducible values can be obtained.

At least, the experimental investigation of the osmotic pressure of colloids points to an enormously high molecular weight, but except under very special conditions cannot be trusted to give a definite value. It should be realized also that any such " molecular weights " refer to a mean value only, for as a rule the particles of sols are not uniform in size, but vary over a wide range. Such a molecular weight has little physical significance ; it might happen that a very small proportion of the particles present actually had an aggregate weight approximating to the mean.

Uniformity of Particles. A system is said to be *polydisperse* when the disperse phase consists of particles of various sizes and unless very special conditions have been operating a colloid is normally polydisperse. By special means it is possible to obtain *isodisperse* systems in which all the particles of the disperse phase are of approximately the same size. The importance of these systems lies in the fact that particles of different degrees of dispersion generally have different stabilities and the properties of ordinary polydisperse colloid systems are a little confused by the fact that the different dispersion fractions react differently towards coagulative and other influences. Although for practical purposes the behaviour of the polydisperse system matters more than that of the more ideal isodisperse system, fundamental studies of primary colloid-chemical processes are best made on the latter simplified system, and it becomes of interest to describe how such systems have been made.

Perrin achieved this object by fractional centrifuging of gum mastic. In the centrifuge the system is put into a tube and whirled at high speed so that the axis of the tube lies in the plane of whirling. Under these conditions suspended particles collect at the end of the tube, and the coarser particles are always thrown out first. The smaller the particles the higher is the velocity of centrifuging required. Perrin collected different fractions of particles by using different speeds; the first treatment brought down the heavier particles, which were then collected and removed, and each successive speed brought down smaller particles. The fractions of particles collected in this way redisperse in water to form sols and each of these sols will contain particles whose size varies between fairly narrow limits.

Zsigmondy's nuclear method (*Z. physikal. Chem.*, 1906, **56**, 65) of preparing gold sols of almost any desired particle size depends on the fact that if a small quantity of an amicronic gold sol is added to a solution in which reduction to gold is taking place (there is generally an interval of a minute or so before the colloid is formed), the supersaturated gold solution condenses on the nuclei provided by the amicrons of the added sol. In practice, the technique is a combination of Zsigmondy's two methods of preparing gold sols (p. 22). A colloidal solution of the finest subdivision is made by reduction with an ethereal phosphorus solution and this is called the nuclear solution. A second mixture of chloroauric acid and potassium carbonate is made and to this formaldehyde is added; but before reduction begins a definite quantity of the nuclear sol is also added to the hot solution. It is of advantage to dilute the solutions recommended on p. 22 so that the reduction without the nuclear liquid would occupy about two minutes. After addition of the nuclear sol the formation of colloidal gold is almost instantaneous and a clear deep red sol is obtained. The size of the particles is regulated by the amount of nuclear sol added, for if few nuclei are presented they develop into comparatively large particles, whilst the addition of larger quantities of nuclei may even leave the particles still microscopic. Further, these larger particles may act as nuclei when added to other solutions in course of reduction and so even larger particles may be produced.

Another principle has been applied by Odén to the preparation of isodisperse sulphur sols from polydisperse sols. In general, the greater the size of the particles the less is the stability of the sol and the more readily is it coagulated by electrolytes. Consequently, in a polydisperse sol particles of different size have different degrees of sensitivity towards coagulating agents, and on adding an electrolyte such as sodium chloride the coarser particles are coagulated preferentially, so that a separation by fractional coagulation is possible. In this way Odén succeeded in separating the polydisperse sol into a series of sulphur sols of different degrees of dispersion and having different properties. This is illustrated by the following table in which the first column shows a concentration (normality) of sodium chloride which is insufficient to precipitate particles of the size concerned, the second column gives the concentration which succeeds in coagulating the particles, and the size of the particles of each fraction is indicated in the third column.

Normality of Sodium Chloride.		Diameter of
Not Coagulated.	Coagulated.	Particles.
0·25	∞	Amicrons
0·20	0·25	,,
0·16	0·20	,,
0·13	0·16	25 $\mu\mu$
0·10	0·13	90 $\mu\mu$
0·07	0·10	140 $\mu\mu$
0	0·07	210 $\mu\mu$

The larger particles of sols may also be removed and the smaller particles retained by means of fractional ultrafiltration. The practical aspect of ultrafiltration is discussed on p. 73, but it may be mentioned here that, although the hardest filter paper allows the passage of colloidal particles, these are stopped by membranes such as parchment which are used for dialysis. Such membranes, which allow crystalloidal particles to pass and retain colloidal particles, can be regarded as ultrafilters. By special methods which, will be described later, ultrafilters of graded porosity can be prepared and, by forcing the colloid through these, particles above a given size can be removed.

Wo. Ostwald and Quast (*Kolloid-Z.*, 1929, **48**, 83, 156) have made measurements of the diffusion coefficients of a number of dyes in mixtures of alcohol and water and have calculated therefrom the sizes of particles in the dyes. In all cases the maximum degree of dispersion was obtained when mixtures of medium composition (40-60 per cent. alcohol) were used as solvent. The radius of the particles may vary with the composition of the solvent mixture to the extent of 1 : 100. In general, the employment of solvent mixtures of varying composition is a simple means of obtaining sols with systematically graded degrees of dispersion.

The Tyndall Effect. The heterogeneity of sols is also demonstrated by an optical effect. A sol may appear to be perfectly clear when viewed by transmitted light, but when a strong beam of light is concentrated in the sol the path of the light is illuminated by a bluish opalescence. This will appear only to a small extent if the refractive indices of the two phases are not very different. This phenomenon is generally called the Tyndall effect, since it was used by him in the examination of fogs and mists. The illumination produced looks like fluorescence, but differs in being polarized, as can readily be observed by means of a Nicol prism. Fluorescent light also has a longer wave length than the incident light.

The effect can be demonstrated simply by condensing a beam of light from the electric arc through a lens so that the cone of condensed light falls in the sol, which is for preference contained in a rectangular trough with glass sides. The Tyndall cone or path of the light rays is illuminated by a bluish light which is well observed in an arsenious sulphide sol. The light cone should also be observed through a mounted Nicol prism, and it will be found that when this is rotated the light appears to be partly or completely extinguished in certain positions, showing that the light is polarized. For comparison, the beam of light should also be condensed in a fluorescent solution such as quinine sulphate, where the Nicol prism will show that the emitted light is not polarized.

The cause of this effect is the scattering of light by the particles of the disperse phase. This is not quite the same as reflection, for the particles are in general smaller than the wave-lengths of visible light ($450-760\mu\mu$) and are therefore unable to reflect the light waves. That is the reason why the particles cannot be seen directly in the microscope. We can imagine the particles being swamped by the light waves, but nevertheless interfering with them and succeeding in diverting some of them. This analogy with water waves is helpful, for at the seaside the waves hurl themselves against the cliffs and can be seen to rebound; the cliff is a large body and corresponds to a particle which can reflect light truly. On the other hand, the waves are not truly reflected from a number of small jagged rocks but are nevertheless impeded and dispersed by them. The analogy must not be taken too far.

Rayleigh has made a study of the intensity of the scattered light and finds that the relations can be expressed by the formula

$$I = k\left(\frac{v^2}{d^2\lambda^4}\right)$$

in which I is the intensity of the light scattered sideways, v is the volume of the particle, d the distance between particle and observer, and λ the wave-length of the scattered light. The formula holds for insulating material and is not applicable to metal particles. It immediately becomes clear from this formula why sols like gamboge, sulphur, arsenious sulphide, etc., show a bluish Tyndall light, for the term λ^4 appears in the denominator, and it follows that small wave-lengths are greatly favoured; consequently the blue colour will predominate. The blue colour of the sky is also explained on this view of the preferential scattering of the small wave-lengths of light.

As the equation contains a volume term, it should help us to determine the size of colloidal particles from measurements of the intensity of scattered light. The difficulty here is that the amount of light scattered depends not only on the size of the particles but also on their shape, and by using different methods of preparation particles of different shapes may be obtained. The study of the amount of light scattered when the sol is under the influence of external forces leads to some information about the shape of the particles, a point to which reference will be made later (p. 46).

The practical use of the Tyndall effect as a method of detecting the presence of colloids is spoiled by the fact that crystalloidal solutions as usually prepared will give the effect to some extent and even distilled water is not quite optically empty. Optically empty liquids can be made by causing a voluminous precipitate to settle in the liquid, during the course of which it carries down with it submicronic particles. For instance, an optically empty solution is produced when sodium hydroxide is added to a solution of zinc sulphate and the precipitate allowed to settle. Nevertheless, in a rough way the Tyndall cone serves as a very useful test.

According to a recent investigation by Schade and Lohfert (*Kolloid-Z.*, 1930, **51**, 65), the purest water, when sufficiently strongly illuminated by the Tyndall cone, does not appear to be optically void and the path of the light is recorded even more clearly by a photographic film. The effect can be intensified by using a source of ultra-violet light and condensing it through a quartz lens, when the Tyndall cone produced in the purest water can readily be photographed. There seems to be no doubt that the phenomenon is due to the scattering of light by the water itself and not by impurities derived from the containing vessel. By repeated distillation of the water a constant end-value for the intensity of the light cone is reached. The intensity increases with lowering of temperature, the rate of change below 20° being greater than over the range 20–90°, and the temperature change is reversible. The addition of minute quantities of pure electrolytes diminishes the amount of light scattered. The curve connecting the intensity of scattered ultra-violet light with temperature agrees closely with the viscosity-temperature curve. The scattering is probably connected with the polymerization of a fraction of the water molecules to large aggregates.

Application of the Ultramicroscope. The final development of the proof of the heterogeneity of colloids has been reached by rendering the particles of the disperse phase visible in the ultramicroscope. The discussion of the practical side of this instrument is reserved for a subsequent chapter (p. 76), and for the present it is only necessary to say that the principle of the ultramicroscope is to examine the Tyndall light under a high-power microscope. Conditions are so arranged that only the light scattered by the particles enters the objective. With great enough magnification the light scattered by each individual particle can be observed as a halo surrounding the particle. Actually the particle itself cannot be seen nor can its diameter be measured directly by this means, but the appearance in the field is a number of bright discs of light, each one representing the light scattered by a separate particle. The ultramicroscope thus reveals the existence of the particles without affording direct visual observation of the particles themselves. The seaside analogy is again helpful here. If one is in a small boat close to a rocky coast such as that of Cornwall it is necessary to keep a look out for submerged rocks which are just below the water level. Before the rocks can be seen their position becomes evident from the curious effect they have on the waves which wash over them. In the same way, in the ultramicroscope we infer the presence of the particles from the disturbing effect they produce on the light waves which swamp them.

An upper limit for the usefulness of the ultramicroscope is imposed by the true reflection of light which occurs when a particle reaches a size comparable with the wave-length of visible light (say $400\mu\mu$), for such a particle shines so brightly in the field as to obscure the effects due to smaller particles. A lower limit occurs at about $5-10\mu\mu$ and with sols containing smaller particles than these the field appears to be uniformly lit. The position of this limit depends very greatly, however, on the difference in the optical properties of the two phases. Obviously, it is an optical heterogeneity that is being recorded in the Tyndall light, and is all the more marked the greater the difference of refractive index between the disperse phase and dispersion medium. As these approach each other in optical properties the scattering becomes less and consequently many colloids appear to be optically empty although the diameter of their particles may be considerably greater than $10\mu\mu$. Other colloids may appear optically empty in spite of a large difference in refractive index of the phases, because the diameter of the particles is less than $5\mu\mu$.

Despite these limitations, the ultramicroscope has proved to be one of the most powerful agencies in the development of the practice and theory of colloids, and the enormous advance in colloid chemistry during the last quarter of a century has been due in no small measure to the invention of this instrument.

The Brownian Movement. The most striking feature of the ultramicroscopic picture is that the bright discs of light are not stationary but are in continual and rapid motion—the Brownian movement, or Brownian movements as they are sometimes called, since there are two kinds of movement, translatory and vibratory. Zsigmondy describes the motion of the particles of gold in a gold sol as resembling " a swarm of dancing gnats in a sunbeam." There is both beauty and aptness in this description, for there is no more orderliness apparent in the motion of these particles than in the erratic movements of the insects.

This remarkable observation revived interest in the phenomenon which had been discovered by Brown in 1827 (*Phil. Mag.*, 1828, **4**, 101). Brown noticed that pollen grains executed continual vibratory movements when immersed in water. This phenomenon is exactly the same as that observed by Zsigmondy in the ultramicroscope, only that the motion of the ultramicroscopic particles is so much more violent as to strike the observer more forcibly with the importance of the phenomenon.

For a long time the movement remained unexplained. Brown, as a botanist, at first thought that it was a phenomenon pertaining to life, but he afterwards convinced himself that any sufficiently finely divided material had this property, which could be observed in an ordinary microscope. He even went so far as to powder a piece of an Egyptian sphinx and show that these undoubtedly lifeless particles exhibited the same movements.

The next question was whether the phenomenon was an inherent property of the physical system or an effect due to outside influences, and this occupied the rest of the nineteenth century. The idea that the movement was due to convection currents was removed, and by working in cellars and in the open fields it was established that external vibrations did not account for the motion. Others ascribed it to local differences in velocity of dissolution, but the movements were found to persist in tiny bubbles enclosed in liquids in certain minerals many thousands of years old, where any such differences must have been equalized during the ages.

The solution came by the application of the kinetic theory to these systems and has been verified by experimental work. Einstein (*Drudes Annalen*, 1905, **21**, 549 ; *Ann. Physik.*, 1906, **19**, 371) made a theoretical investigation on the assumption that there is no difference between true molecules and larger particles suspended in the same medium, the particles behaving as though they were true gas molecules with normal kinetic energy but a much shorter free path. The result of Einstein's reasoning is the formula

$$A = \sqrt{t} \sqrt{\frac{RT}{N} \cdot \frac{1}{3\pi\eta r}}$$

in which A is the amplitude of displacement of the particles, t is the corresponding time or duration of the displacement, R is the gas constant, T the absolute temperature, N the Avogadro constant or number of molecules in one gram molecule, η the viscosity of the medium, and r the mean radius of the particles.

Simultaneously, von Smoluchowski (*Ann. Physik.*, 1906, **21**,

756) arrived at the same formula by different reasoning, with the exception that another integer is introduced.

At the same time as these theoretical investigations, Svedberg (*Z. Elektrochem*, 1906, **12**, 853, 909) was carrying out experimental determinations of the Brownian movement by means of the ultramicroscope. He found that when the sol under observation was allowed to flow slowly across the field of vision the Brownian movements were superimposed on the general drift of the particles and consequently gave the appearance of luminous wave-like curves. Eventually a refined technique for photographing these waves was developed and from the record the average amplitude of the displacements and the wave-lengths could be read directly. Knowing the velocity of flow of the sol, the corresponding time interval could be calculated. The following table gives results obtained in experiments with platinum particles of about $25\mu\mu$ radius at 19° in various dispersion media. The symbols have the same significance as in the Einstein formula above.

Dispersion Medium	$\eta \times 10^1$	l (in μ)	t (in sec.)	$l/t \times 10^2$	$l\eta \times 10^2$
Acetone	3 2	6 2	0 016	3 9	2 0
Ethyl acetate	4·6	3 9	0 014	2 8	1·8
Amyl acetate	5·9	2·9	0·013	2·2	1·7
Water	10 2	2·1	0·0065	3 2	2·1
N-propyl alcohol	22 6	1·3	0·0045	2 9	2·9

From these figures Svedberg deduced two important experimental results, viz. $A/E =$ const. and $A\eta =$ const., or expressed in words, the amplitude is proportional to the periodic time and inversely proportional to the viscosity. The two expressions may be united by multiplying them together into the expression

$$A^2 = \text{const.} \times \frac{t}{\eta}.$$

Although at the time of carrying out the experiments, Svedberg was not acquainted with the theoretical investigations of Einstein and of von Smoluchowski, it soon became apparent that this work is an experimental verification of their reasoning, for when T and r are kept constant Einstein's expression reduces to this identical form, $A^2 =$ const. $\times t/\eta$.

Consequently it must be regarded as established that the suspended particles have the same kinetic energy as gas molecules and are subject to the same laws; they exert an osmotic pressure on a membrane impermeable to them, the value depending on the temperature and the number of particles in unit volume. This osmotic pressure is very small and generally defies measurement, since from the molecular kinetic point of view colloidal solutions are extremely dilute.

According to the kinetic theory of gases, the temperature is determined by the kinetic energy of the molecules, which is given by the expression $\frac{1}{2} m v^2$, where m is the mass and v the velocity. If this kinetic energy is equally distributed between the molecules of the dispersion medium and the particles of disperse phase it follows that the mean velocity of the colloidal particles in Brownian motion must be correspondingly less than molecules just as the mass of the particles is greater.

The Brownian movement is therefore due to bombardment of the colloidal particles by the molecules of the dispersion medium. As the particle increases in size, the probability of unequal bombardment diminishes and eventually the collisions on different sides equalize each other, and in practice the Brownian movement becomes imperceptible when the particles reach a diameter of about 3μ. As the size of the particle is reduced, the probability of unequal bombardment rapidly increases and consequently the Brownian movement becomes more and more violent.

It will be clear that the Brownian movement is no mere curiosity, but gives an insight into molecular reality. It does more, for since the formula of Einstein is established on an experimental basis it enables us to calculate N, the Avogadro constant. We may consider the classical work of Perrin (*Kolloidchem Beih.*, 1910, **1**, 221) on this subject.

In these experiments Perrin employed disperse systems of gamboge and mastic, which were carefully prepared and separated by fractional centrifuging into systems having particles of uniform size, ranging from 0·212 to 11·5μ. These particles are visible in the ordinary microscope. The radius of the particles was measured by methods which will be discussed later, the viscosity and temperature of the liquid were known, and by measuring under the microscope the amplitude of displacement of a particle and the corresponding periodic time, all the data were obtained for determining N by substitution in the Einstein formula. Results are shown in the following table.

COLLOID SYSTEMS

Material.	Particle Radius (μ).	N
Gamboge	0·50	$6·6 \times 10^{23}$
,,	0·212	$7·3 \times 10^{23}$
Mastic	5·50	$7·8 \times 10^{23}$
,,	0·52	$7·25 \times 10^{23}$
,,	0·367	$6·9 \times 10^{23}$

The values of N obtained agree very satisfactorily with those deduced from entirely different methods, and the agreement among the values is extraordinarily striking when it is remembered that a particle of radius $5·5\mu$ has 17,000 times the volume of one with a radius of $0·212\mu$. This experimental work also affords useful confirmation that the theoretical foundations both of the kinetic theory of gases and of its application to systems of large particles are on right lines.

Another consequence of the theory is that the particles in a disperse system should arrange themselves under the influence of gravity in the same way as the molecules of a gas. The atmosphere surrounding the earth is known to be less dense at high altitudes and a simple formula has been evolved which connects the change of density with the height. Perrin carried out further experiments with his isodisperse gamboge and mastic systems in order to try and confirm this same relation for these cases.

His method was to place the suspension of particles under the microscope and focus the latter at different heights and count the number of particles visible in the field at each height. The variation in the number of particles with the height was found to agree remarkably well with that calculated on the assumptions given above. The following table shows some of the results, the number of particles observed at a height of 5μ being arbitrarily fixed at 100 and the other numbers in proportion.

Height in μ.	No of Particles (obs.).	No. of Particles (calc.).
5	100	100
15	43	45
25	22	21
35	10	9·4

These results were obtained on the examination of a very thin layer at the surface of the liquid. Obviously, it is quite impossible that so great a change of concentration with height could exist throughout the whole liquid, for it can be calculated from these data that the concentration of a silver sol at a given layer would be 5,000 times that at another level 1 cm. higher, and it is quite evident in any case that the concentration of a sol in stable equilibrium with gravity is practically uniform throughout almost the entire depth.

The distribution of particles under gravity in accordance with the kinetic theory only holds then for the particles in the surface of the liquid. It has been pointed out that the reasoning involves the assumption of the validity of Boyle's law, and this is untenable when the volume of the particles is large. More recently, extensive studies of the theory and practice of sedimentation equilibrium have been made by Svedberg (*Handbuch der biologischen Arbeitsmethoden*, Teil B, Heft iv, Berlin, 1927).

Determination of Particle Size. The size of the particles cannot be measured directly in the ultramicroscope on a micrometer eyepiece because the true image of the particle does not appear, but only a halo of light surrounding the particle and indicating its position. The size can be determined by an indirect means, however, using the slit ultramicroscope (p. 76). In this instrument a small volume of the sol is brilliantly illuminated and a small portion of this delimited in the microscope field. The volume actually under observation in the ultramicroscope is of the order of cubic μ and can be measured. The number of particles to be seen in the field are counted at different times and an average is taken. This gives the average number of particles present in a known very small volume of the sol. The total weight of disperse phase in say 100 c.c. of the sol can be found by evaporating the sol to dryness and weighing the residue. Consequently, for a given volume of sol we know both the total weight of the particles and the number of particles. This gives directly the average weight of each particle. From this weight the dimensions of the particle are readily found by assuming that the density of the particle is the same as in the massive state and that the particles are spherical or have some other simple geometrical form. Probably, neither of these assumptions is quite correct as a rule, but there is no doubt that the results obtained are pretty close to the truth.

Although ultrafiltration gives only a one-sided index to the particle size it may be useful in showing that the particles are below a certain size. The size of the pores of ultrafilters can be determined by methods which are described on p. 75. However, the ultrafilter does not act purely as a mechanical sieve, and particles of some substances which are small enough to pass through the pores are retained by the ultrafilter, probably by electrical forces operating at the boundary. At the same time, if the particles pass through the ultrafilter it is reasonable to suppose that their diameter is less than that of the pores, unless they are easily deformed, and to this extent ultrafiltration gives an index of particle size.

Particles too small to be observed in the ultramicroscope can sometimes be measured by Zsigmondy's ingenious nuclear device. It is possible to "develop" gold amicrons, or deposit gold on them until they become big enough to be seen and counted in the ultramicroscope (cf. p. 34). Knowing the mass of disperse phase in the original amicronic sol the size of the original particles may then be calculated. It is important that conditions be such that no more nuclei are formed. In this way it has been possible to measure the radius of particles down to about $1\mu\mu$.

Particles of larger size can be measured by determining their rate of fall under gravity in a liquid of known viscosity. This occurs with a definite velocity which is expressed in Stokes' law (p. 89). Odén (*Proc. Roy. Soc.*, *Edinburgh*, 1916, **32**, 219) has developed a very ingenious automatic method of recording the amount of sedimentation with time of accumulation on a plate immersed in the disperse system and suspended on a balance. The method has since been modified by Svedberg (*J. Amer. Chem. Soc.*, 1923, **45**, 943).

Other general methods available for determining the size of particles are measurements of osmotic pressure, diffusion, light absorption, and intensity of scattered light, but the existence of numerous complicating factors in either the measurement or interpretation of these has so far made them of little service as applied to colloid systems.

Wo. Ostwald and von Buzágh (*Kolloid-Z.*, 1929, **47**, 314) have pointed out an interesting relation between the dispersity of a substance and its chemical composition. The composition of the elementary cell of a crystal as disclosed by X-ray examination differs from the stoicheimetric composition of large crystals, the two merging into one another as the size of the crystal increases. This progression leads to a convergence of the composition in crystals having a length of about 10^{-6} to 10^{-5} cm., i.e. typical colloidal dimensions. It follows that the composition of the colloidal particles in sols and highly disperse precipitates can vary with the size of the particles. For example, a highly dispersed calcium fluoride sol prepared by a condensation method may be richer in calcium ions than a similar sol of lower dispersity.

The Shape and Structure of the Particles. The shape of colloid particles cannot be discerned directly in the ultramicroscope, for the true image does not appear and any fairly regular shape such as sphere, cube, or octahedron is observed as a disc of scattered light. If, however, the particles are abnormally developed in one direction so as to be rod-like, the discs of light observed are correspondingly elongated. If the particles are developed along two rectangular axes they are plate-like, and such particles exhibit a peculiar twinkling effect in the ultramicroscope. This is because they are always on the move and more light is scattered by the side of the platelet than by the thin edge.

Even without ultramicroscopical examination some information about the shape of the particles may be gained in certain cases by a macroscopic optical effect. When sols containing fairly large rod-shaped particles are stirred the lines of flow in the sol are marked by dark and light streaks. This phenomenon is also shown by coarse suspensions of long-shaped particles, such as those of asbestos, and is caused by the orientation of the particles in the direction of flow. Obviously in the local current, where all the particles are broadside on, there will be much greater reflection of light than in the body of liquid, where the particles are arranged at haphazard.

Freundlich has made a quantitative study of this effect by giving the liquid a constant motion and measuring the intensity of light scattered in different directions when the sol is illuminated by a beam of linearly polarized light. Then the intensity of the scattered light in a certain direction depends on the orientation of the particles, the position of the electrical vector, and the direction of the incident beam.

In this way it was found that the particles of red gold sols, silver, platinum, arsenious sulphide, and mastic are approximately spherical, that platelets occur in blue gold sols and aged ferric hydroxide sols, whilst sols of vanadium pentonide which have been kept for some months, tungstic acid, and certain dyes (e.g. aniline blue) have rod-like particles. One of the most interesting of these observations is the ageing effect observed with certain sols, some of which have roughly spherical particles when freshly prepared, but change on keeping to another shape, often rod-like, with a corresponding change in the optical properties. This is an important observation, because many of the properties of colloids vary with time and are quite different for different preparations of the same substance kept for varying lengths of time. This feature has come to be regarded as so characteristic that it is usual in many researches to state the age of the colloid.

Another interesting incident is the property of all pure red gold sols to turn blue during coagulation. This subject has received a large amount of study, into which it is not proposed to go here, but it may be pointed out that it is impossible to explain the colour change on the grounds of an increase in size, for the change takes place in amicronic gold sols, where the final blue particle may still be much smaller than the primary particle of a less highly dispersed red gold sol. It is concluded mainly from optical investigations that the change of colour is due to the loose union of the primary gold particles into aggregates which are still of colloidal size. Zsigmondy draws the important conclusion that in this coagulation there is little decrease in total surface, whatever decrease does take place being confined to the edges that touch.

Zocher and Jacobsohn (*Kolloidchem. Beih.*, 1929, **28**, 167) have given the term " tactosols " to sols containing non-spherical particles which have the property of spontaneously arranging themselves in parallel order. This property has been studied with sols of vanadium pentoxide, benzopurpurin, ferric hydroxide, tungsten trioxide, and chrysophenin. Vanadium pentoxide separates on ageing into a concentrated anisotropic phase—the tactosol—which later precipitates out, and a dilute isotropic phase, which is termed the atactosol. When an electric field is applied, the particles arrange themselves with their long axes parallel to the direction of the current, but if an alternating current is applied the orientation of the particles is perpendicular to the current. In a magnetic field the particles are arranged parallel to the lines of force. Tactoids are readily produced by cooling a 2 per cent. boiling sol of benzopurpurin—4B, but benzopurpurin—6B exhibits the effect better, even in a 1 per cent. sol. Tactosols of tungsten trioxide have disc-like particles bearing a negative charge and consisting of a number of parallel platelets having a constant " period " or distance apart. The tactoids of chrysophenin consist of long, lamellar, negative anisotropic crystals.

Many sols, which contain elongated particles, become doubly refracting when stirred or when placed in an electric or magnetic field. The cause of this is that either in the stream line or in the applied field the particles are oriented with their optical axes in one direction so that they act as one crystal unit, and being crystalline and doubly-refracting they confer this optical property on the sol as a whole. The property of becoming birefringent in a magnetic field can be taken as an indication of dissymetry in shape of the particles and has been used in investigations along these lines. By this means it has been shown that ordinary sulphur sols prepared by chemical reactions give no double refraction, whilst sulphur sols prepared by grinding sulphur crystals are doubly-refractive. This probably means that particles of the latter sols are crystalline and dissymmetrical in shape and that particles of the former sols are spherical and perhaps amorphous droplets of sulphur.

This brings up a point which was once the cause of a good deal of diversity of opinion ; the question whether the ultimate particles of colloids are crystalline or amorphous. Largely because of Graham's first differentiation of colloids from crystalloids and because of the properties of such typical colloids as gelatin in the dry state it was assumed that colloids were composed of amorphous material. It is not surprising, therefore, that the optical properties which have just been mentioned were regarded with no little interest. This was before the development of the present methods of investigating the crystal structure of particles by X-ray spectrographic analysis. Details of these methods are given in text-books of physics and their application constitutes a very great part of modern research.

X-ray analysis of sols of the noble metals and many other substances shows quite definitely that the colloidal particle is generally crystalline. This does not preclude the existence of liquid particles, as in emulsions, or of solid amorphous particles, as in sulphur, but in general the colloidal particles of substances which are normally obtained in a crystalline state are themselves crystalline. Regarded from the point of view of the present age, and uninfluenced by older views on the nature of colloids we should not expect them to be otherwise. Obviously, the loose differentiation sometimes made between crystalline and colloidal material has no foundation.

The terms "liquid," "solid," and "amorphous" are still used rather indefinitely at times. The so-called amorphous precipitate generally consists of an assemblage of tiny crystals; it would be correct to term it microcrystalline. Similarly, the particles of sols of the colloidal gold type may be called ultramicrocrystalline. The crystalline state is characterized by an orderly arrangement of the units of the structure in what is called a space-lattice. A disorderly arrangement can be called amorphous and this may be either liquid or solid in accordance with the mechanical properties of the substance. It does not seem correct to call glasses supercooled liquids, but rather amorphous solids, although considerable diversity of opinion occurs in matters of this sort. In the same way, an orderly arrangement of molecules may occur in a liquid and then we have crystalline liquids or liquid crystals.

One thing to be remembered is that if a microcrystalline precipitate is compressed, the compact mass formed must be regarded as being amorphous in part, for, where the boundaries of the crystals touch, the adjacent particles cannot form a normal part of the lattice of both crystallites. In such a compact mass the amount of this intergranular boundary is exceedingly high. These considerations are very important in problems of metallurgy and mineralogy, but cannot be further discussed here.

Many of the substances which are regarded as typically amorphous, such as rubber and cellulose, give an X-ray interference diagram when stretched. In the unstretched state the usual "amorphous ring" is obtained. This indicates an orientation of particles on stretching, giving a close approach to a space-lattice and crystalline properties. Indeed, some of the stretched materials acquire a definite fracture in certain directions.

Trillat (*Compt. Rend.*, 1929, **188**, 1246) has made an X-ray examination of stretched films of cellulose, cellulose nitrate, and cellulose acetate and finds that the diagrams become more and more like those of crystalline substances as the stretching increases. The passage from the gel state appears to occur gradually, the final condition being that of pseudo-crystallization in which the molecules occupy the same positions as in the crystal lattice, but are not in perfect alignment. It is pointed out that modification of chemical properties may result as a consequence of change in orientation of active groups.

Similar experiments carried out by Mark and von Susich (*Kolloid-Z.*, 1928, **46**, 11) show that when caoutchouc is stretched an orientation of the particles takes place, not only in the direction of stretching, but also in a perpendicular direction. The stretched caoutchouc belongs to the rhombic crystal system, and there are eight isoprene residues in the elementary cell. Von Susich (*Naturwiss*, 1930, **18**, 915) has recently made an interesting development of this work. Unstretched natural caoutchouc gives an X-ray interference diagram below 35–38°, but not above that temperature. With increasing degree of stretching the temperature at which the substance is truly amorphous rises. By plotting the degree of stretching against the temperature at which an X-ray pattern ceases to be obtained, a " melting curve," showing the transition from crystalline to amorphous caoutchouc, is obtained.

CHAPTER V
ELECTRICAL PROPERTIES OF COLLOIDS

Colloid particles are electrically charged, a fact which was firmly established by Linder and Picton in 1892 (*J. Chem. Soc.*, 1892, **61**, 148 ; 1897, **71**, 568), who showed that when the sol is placed in an electric field the particles of disperse phase move towards one or other of the electrodes.

Cataphoresis. This electrical migration of colloid particles is called *cataphoresis* and can readily be demonstrated in the apparatus used by Burton (*Phil. Mag.*, 1906, **11**, 436), who carried out some important investigations in this connexion. The apparatus is shown in Fig. 2. Part of the U-tube is filled with dispersion medium and then some of the sol is run in slowly through the funnel so as to form a layer under the pure dispersion medium. A potential difference is applied across platinum electrodes immersed in the dispersion medium. If the sol is coloured the rate of motion of the boundary of the sol can be observed directly; if not, a modified form of apparatus is used and the direction and amount of transport are found by analysis.

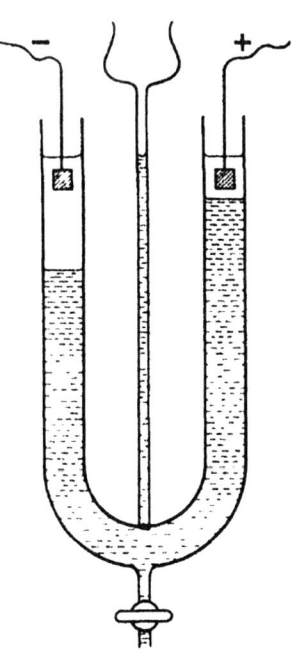

FIG 2 —Cataphoresis.

In most cases the best determinations of cataphoretic migration velocity are made by microscopical or ultramicroscopical methods.

A small quantity of the sol is placed in a tiny cell with platinum electrodes and a portion of the sol is observed under the ultra-microscope. Then the rate of movement of individual particles under a known potential gradient can be followed directly. Further details are given on p. 78.

Tested in this way, it is found that the particles of some sols move towards the anode, whilst others migrate towards the cathode. Obviously, sols of the first type contain negatively charged particles and sols of the second type contain positively charged particles. The majority of sols migrate to the anode or are negatively charged, and typical of these are sols of metals and metallic sulphides; sols of metallic hydroxides as a rule are positively charged. The following list gives an idea of the sign of charge associated with typical colloids, but it must be remembered that under certain conditions the sign of charge may be reversed.

Negatively charged sols.	Positively charged sols
Gold, silver, platinum.	Ferric hydroxide
Sulphur	Aluminium hydroxide
Arsenious sulphide	Chromium hydroxide
Cupric sulphide	Cadmium hydroxide
Cadmium sulphide	Titanic acid
Mastic	Ceric hydroxide
Gamboge	Thorium oxide
Silicic acid	Zirconium oxide
Stannic acid	Basic dyestuffs
Molybdenum blue	
Tungsten blue	
Vanadium pentoxide	
Acid dyestuffs	

When the colloidal particles reach the electrode they lose their charge and generally coagulate into coarse particles. This fact suggests a close relation between the charge on the particles and the stability of the sol, a matter which will be discussed in some detail later. That there is a connexion between electric charge, stability, and the presence of ions is shown by some work carried out by Burton (*loc. cit.*), who studied the influence of an added aluminium salt on the cataphoretic velocity of the particles of a gold sol and showed how, at the same time, the stability of the sol changes. These results are reproduced in the following table.

Mg. Al per litre	Velocity in μ per volt per cm. per sec.	Stability
0	330 (towards anode)	Stable indefinitely
0·19	171 ,, ,,	Flocculated after 4 hours
—	0	Immediate flocculation
0·38	17 (towards cathode)	Flocculated after 4 hours
0·63	135 ,, ,,	Incompletely flocculated after 4 days

The chief point to notice is that with increasing addition of small quantities of aluminium chloride the migration velocity decreases to zero and later increases because of a reversal of charge of the particles. The stability is also at its lowest when the particles are uncharged and the stability increases with the charge of the particle, whether negative or positive.

These experiments aroused interest by confirming some previous results of Hardy (*Z. physikal. Chem.*, 1900, 33, 385), which had remained unaccepted previously. Hardy found that a sol of denatured egg albumin varied in sign according to whether acid or alkali was present. The sol was stable both in presence of acid and of alkali, but was positively charged in the former case and negatively charged in the latter. In the absence of acid or alkali the sol flocculated. Hardy therefore expressed the opinion that a colloidal solution can be stable only when the particles possess an electric charge and that, when deprived of this charge, the particles unite to form coarse aggregates which eventually precipitate.

It has since been firmly established that the presence of electrolytes has a powerful effect on the potential difference between the particle and the dispersion medium and that this is closely connected with the stability of the sol. A particularly important observation is that the action of the ions increases very greatly with the valency. The following table contains some results obtained by Freundlich, and gives the concentration necessary to lower the potential between glass particles and water from 0·089 to 0·039 volts. It shows how enormously the effect of the ion varies with its valency. It will be shown later that the valency of the ion affects the stability of the sol in the same way.

Electrolyte.	Millimols per litre.
KCl	25
$BaCl_2$	0·87
$AlCl_3$	0·02
$ThCl_4$	0·015

Electro-osmosis. Since the particles of disperse phase move relatively to the dispersion medium under the influence of an electric field, it follows that if the particles could be maintained stationary the dispersion medium would move when the field was applied. In practice this state of affairs can be realized by making the substance of the disperse phase into a porous diaphragm and fixing this in position in a tube containing the dispersion medium. When a potential difference is set across the ends of the tube the dispersion medium moves towards one or other of the electrodes, the direction being opposite to that which the diaphragm would follow if it were free to move. This movement of the dispersion medium under the influence of the electric field is known as *electrical endosmose* or *electro-osmosis*.

This phenomenon is the counterpart of cataphoresis and is modified in a similar way by the addition of acids, alkalis, and neutral salts. In general, negative diaphragms become more negative in alkaline solutions, and in acid solutions they become less negative, electrically neutral, and finally positive. Correspondingly, positive diaphragms become more positive in acid solutions, whilst with increasing concentration of alkali they become less positive, electrically neutral, and finally negative. In the case of neutral salts, the valency of the ion is again of very great importance. These changes can be followed by measuring the amount of liquid transported in a given time at a given applied potential. These phenomena are more easily investigated in experiments on electro-osmosis than in cataphoretic experiments, for there are less disturbing factors. When electrolytes are added to sols, partial or complete coagulation usually occurs, but this inconvenience is removed by using the solid phase as a fixed diaphragm.

In experiments of this type, where the electric charge at an interface may be reversed in sign, there must be some intermediate position where there is no difference of potential between the two phases. This position is called the *isoelectric point*, and where acids and alkalis are concerned in determining the sign of charge of the particles the isoelectric point occurs at a definite hydrogen-ion concentration, the exact determination of which is a matter of some importance in the study of certain types of colloids. At the isoelectric point no migration occurs in the electric field and as a rule sols have a maximum of instability and

COLLOIDS

generally precipitate out. There are a few exceptional cases, where the maximum instability does not correspond to the isoelectric point of the sol, but these are not understood at present.

Streaming Potential. In electro-osmosis a liquid flow across a solid surface or diaphragm is set up by an applied electric field; conversely, if the liquid is forced through the diaphragm by pressure in absence of applied electric field a difference of potential is set up. This is known as the streaming potential. It may be simply measured by forcing water through a capillary tube of the substance under examination and determining the difference of potential between electrodes placed at each end of the tube.

Determination of the Charge. Qualitatively, a cataphoretic experiment shows at once the sign of charge on the particles, but methods have been worked out for finding the interfacial potential difference from measurements of the velocity of cataphoresis, the velocity of electrical endosmose, or the streaming potential.

The equation giving the relation between the potential of the particle relative to the liquid (ζ) and the velocity of cataphoretic migration (u) is

$$\zeta = \frac{4\pi\eta}{HD} \cdot u$$

where η is the viscosity of the liquid, H is the potential gradient in volts per cm., and D is the dielectric constant of the medium.

The following equation connects the potential with the volume (v) of liquid transported per second by electro-osmosis.

$$\zeta = \frac{4\pi}{iD} \cdot \eta k v$$

Here, i denotes the current strength, k the specific conductance of the liquid, D the dielectric constant of the medium, and η the viscosity.

The potential of the particles may be calculated from measurements of the streaming potential by means of the equation

$$\zeta = \frac{4\pi}{D}\eta \cdot k\frac{E}{P}$$

where E is the streaming potential, P is the pressure under which the liquid is forced, and the other terms have their former significance.

The potential is always small and is often expressed in millivolts.

Calculations have been made of the charge on particles of gold hydrosols of different radius in terms of electrons and the following table is illuminating.

Radius in μμ.	Charge on particle in No. of electrons.
1	6
2	14
10	120
24	550
100	8,550
240	47,000

It shows that the charge is high and that the mobility of colloid particles is of the same order of magnitude as those obtained for ions. It follows that sols must possess a certain conductivity. This is small compared with that of normal solutions of electrolytes because, although the mobility of the particles is comparable with that of ions, colloids are extremely dilute solutions as regarded from the molecular kinetic point of view. Consequently there is always grave doubt in the interpretation of conductivity values obtained experimentally with sols, because in most cases there are electrolytes present and it is difficult to know what part of the measured conductivity is due to the cataphoretic migration of the particles. It is always found that the conductivity of the sol decreases as the sol is purified.

Some colloids are quite definitely ionized, many of them to a large extent. Soap solutions and many dyes are typical colloidal electrolytes. Many of these substances ionize so as to give one simple ion such as the sodium ion and another complex ion of colloidal dimensions. In substances of this type, occurring mainly among the organic colloids, the conductivity values obtained are much higher and are often of great importance.

The Electric Double Layer. The facts which have been discussed in the preceding section often give rise to the idea that the sol as a whole is an electrically charged body and can be represented as a collection of statically charged spheres. This is very wide of the truth; there is no external charge on a colloid and no energy can be obtained from it externally by this means alone. If so, it would be possible to charge or discharge an electroscope by bringing it into contact with a sol. The charge resides on the surface of each particle and is compensated by the equal and opposite charge on the surface layer of liquid immediately in contact with the particle; hence, the system as a whole appears to be uncharged. This idea of the electric double layer was originally due to Helmholtz and is schematically represented in Fig. 3. The electrokinetic phenomena described in the preceding section occur whenever there is displacement in the electric double layer caused by the existence of slip between the two coatings of this double layer. When an electric field is applied, the coatings move in opposite directions, tending to set up an opposing electromotive force. This tendency is compensated, however, by the intermicellar liquid, which is a peptizing electrolyte and hence a conductor of electricity, so that the net result is the movement of the particle and its double layer towards the electrode. The colloid particle plus its electric double layer is often called the *micelle*.

This conception of the electric double layer involves the supposition that the potential gradient at the interface between the two phases is abrupt and that oppositely charged layers are at a molecular distance from each other. The Helmholtz idea has been replaced by the conception of a

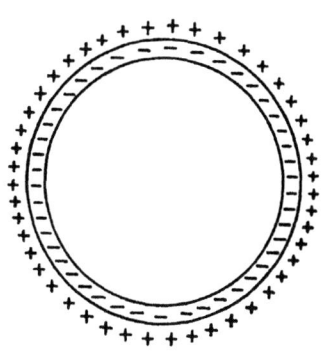

FIG 3 —Helmholtz Electric Double Layer.

diffuse double layer, mainly due to Gouy (*J. Physique*, 1910, **9**, 457), according to which the potential gradient is not sudden but is diffused over a short distance. The double layer is supposed to be built up of two diffuse layers of ions, the concentration of say the negative ions decreasing and the concentration of the positive ions increasing with increasing distance from the centre. Gouy calculated the thickness of the double layer that would be equivalent to such a mixed atmosphere of ions and found it to decrease as the concentration of the electrolyte increases. For ordinary sols the thickness of the double layer is calculated to be a few $\mu\mu$.

That the diffuse double layer is a better representation of the facts than the Helmholtz idea is supported by some arguments brought forward by Freundlich and Rona (*Sitz. Preusz. Akad.*

Wiss. 1920, **20**, 397), who pointed out that the electrokinetic potential measured by cataphoretic and electro-osmotic methods is less than the static boundary potential. Suppose, for instance, we blow a very thin bulb at the end of a glass tube, immerse it in a solution and fill the inner part with an electrolyte. Connect the inner electrolyte to an earthed electrometer by means of a platinum wire and connect the solution through a calomel electrode to earth and we have a combination which enables us to measure the potential difference between the interior of the glass and the outer liquid. This is an electrostatic potential, ε. The electrokinetic potential, ζ, can be determined for the same type of glass by measuring the streaming potential in a capillary and using the formula given on p. 55. It is found that ζ is less than ε. Another difference is that the electrostatic potential is affected very little by the presence of ions, whilst the electrokinetic potential varies very greatly with the kind and quantity of ions present. The valency of the ion has no influence on the electrostatic potential, but is all-important in determining the electrokinetic potential.

The difference between these values is that one is always determined in connexion with phenomena of motion, and the liquid immediately contiguous to the wall of the solid phase does not move because the latter keeps a thin layer of liquid tightly bound by adsorption. Consequently, the electrokinetic potential is really a measure of the potential difference between the moving and the stationary part of the liquid, whilst the electrostatic potential measures the *total* difference of potential between the solid wall and the liquid. The stability of colloids is governed by the electrokinetic potential, not by the electrostatic potential.

The Origin of the Charge. Although for most systems we can give a reasonable explanation for the existence of the electric double layer, it is not at present possible to give a general rule which is applicable to all types of systems including hydrosols and organosols. Perhaps this is because there are several mechanisms by which the charge can come into being and any of them may operate, depending on the particular circumstances.

In the early days there arose the question whether the charge resembled the frictional electricity brought about by the contact of such substances as glass and silk, or whether the colloid particles became charged by the preferential adsorption of either positive or negative ions of some kind from the dispersion medium. In general, the interface between two dissimilar substances is the seat of an electric potential difference and the rule is that the substance with the higher dielectric constant acquires a positive charge. Since water has an exceptionally high dielectric constant it has been argued that this explains why the majority of hydrosols contain negatively charged particles. On the other hand, the marked influence of ions and especially of the heavily-charged polyvalent ions on the electrokinetic potential and the stability of colloids points to an ionic source of the charge, especially considering that the sign of the charge may be reversed by the addition of certain ions. Moreover, as will be apparent later, we know quite definitely that ions are adsorbed when colloids are coagulated by electrolytes, for analysis shows the coagulum to contain one of the ions of the precipitating electrolyte.

Some twenty years ago, the view adopted in respect of hydrosols was that a neutral wall in contact with water acquired a negative charge, since it had a preference for adsorbing the hydroxyl ions of water; a surface of decidedly basic material such as ferric hydroxide was supposed to have a preference for adsorbing hydrogen ions and hence gave a positively charged colloid; a substance of acidic character adsorbed hydroxyl ions preferentially and thus received a negative charge. There is in some cases independent experimental evidence to show that one of the electrolytic dissociation products of water may be more strongly adsorbed than the other. For example, a semipermeable membrane is formed at the boundary of solutions of aluminium salts and of ammonia, which is evidence that the precipitate is impermeable to the hydroxyl ion. The behaviour of aluminium anodes in certain solutions also shows that whilst even a large electromotive force fails to drive hydroxyl ions through a film of aluminium hydroxide, such a film is readily permeated by hydrogen ions. Consequently, particles of aluminium hydroxide suspended in pure water may be expected to take up more hydrogen ions than hydroxyl ions, so giving the particle a positive charge, while the corresponding hydroxyl ions form a negatively charged atmosphere outside the particle.

In 1912, Freundlich and Elissafoff (*Z. physikal. Chem.*, 1912, **79**, 385) suggested that the material of the particle itself might give rise to the ions which participate in the formation of the double layer, and this idea has since been developed in many directions with marked success. The generalization of the view that the lining of the double layer is composed of ions from the material of the particle itself appears to be reasonable in the case of ionogenic substances such as glass, metallic hydroxides and sulphides, salts, etc., but difficulties arise with typically covalent compounds such as naphthalene. Some modern views of the capillary electrical properties of these substances suggest that these difficulties may be overcome, but on the other hand the origin of the charge may be ascribable to different causes in different instances.

One might suppose right away that the electric double layer associated with the particles of a metal hydrosol is identical with that which enters into the well-known Nernst theory of solution tension of these metals, the double layer being formed by the metals going into solution. If that were so, the electromotive series might reasonably be expected to play some part in the order of capillary electrical phenomena, but it does not. It is also easy to adduce arguments showing that the electric double layer of a silver particle, for instance, does not consist exclusively of silver ions sent into solution by the silver. The presence of hydroxyl ions favours the formation of negatively charged metallic hydrosols, but ammonia is actually harmful to the preparation of a silver sol, and since silver oxide is easily soluble in ammonia it is difficult to escape from the conclusion that silver oxide or some oxidic compound of silver enters into the formation of the double layer.

More definite information may be gained from studies carried out by Zsigmondy and his pupils on the peptization of a stannic oxide sol. When stannic chloride is poured into water a gelatinous precipitate of hydrated stannic oxide is found and the hydrochloric acid formed simultaneously by hydrolysis may be removed by repeatedly decanting the supernatant liquid. The precipitate is brought into colloidal solution by adding either an acid or a base ; in the first case a positively charged sol is produced and in the second case a negative sol.

So far the phenomena could be explained on the older view that hydrogen ions were adsorbed preferentially in the one case and hydroxyl ions in the other, but experiments on the coagulation of the stannic oxide sol by electrolytes point to a different explanation.

A number of sols of stannic oxide were prepared by peptization with potassium hydroxide so that the ratio SnO_2/KOH was different in each case, and the minimum concentration required to coagulate each sol was determined for a number of electrolytes. The results are given in the following table, where the coagulation values are expressed in milliequivalents per 10 c.c.

Ratio SnO_2/KOH	2	10	25	50	100
NaCl	1·8	1 7	0 34	0·26	0 14
NaNO₃	1 9	1·5	0·30	0 28	0·14
Na₂SO₄	1 8	1 7	0 32	0 28	0 14
NaH citrate	2·3	2	0 40	0 52	0 50
HCl	0 33	0 07	0 025	0·0135	0 007
CaCl₂	0 33	0 075	0 022	0 0135	0 007
BaCl₂	0 35	0·065	0 022	0 0130	0·007
AlCl₃	0 33	0 07	0·025	0 0135	0 007
Al(NO₃)₃	0 33	0 075	0 024	0 0140	0 007
AgNO₃	—	—	0 025	0 0180	0·009
Alkali in 10 c.c.	0·333	0 065	0 026	0·013	0 0064

The features of this table are that for each sol the coagulation values (or minimal quantity required for coagulation) of the various sodium salts are nearly equal and form one group, whilst such dissimilar ions as hydrogen, calcium, barium, aluminium, and silver have almost the same coagulation values and form another quite distinct group. The most surprising feature appears on the last line of the table, where it is shown that the amount of these ions required for coagulation is chemically equivalent to the amount of peptizing alkali present.

From these data, Zsigmondy reached a very important conclusion concerning the nature of the double layer in this system. He deduces that peptization does not take place by direct adsorption of hydroxyl ions but that the added potassium hydroxide reacts at the surface of the stannic oxide particles to form potassium stannate, which dissociates forming potassium ions and stannate ions which are the components of the electric double layer. This scheme is represented in Fig. 4 (a), where, however, no attempt is made to indicate the diffuse layer.

On this view the coagulating effect of the various cations is easily understood, for the ions in the upper part of the table form soluble stannates, whilst those in the lower part form insoluble stannates. Consequently the particle is precipitated when the boundary layer is destroyed through the dissociated potassium stannate being converted into an insoluble stannate. In the case of hydrochloric acid, stannic acid is the product and this is insoluble in water. The chemical equivalence between the amount of coagulating electrolyte and the amount of peptizing alkali is immediately understood.

It must not be imagined, however, that such a chemical equivalence is always observed or that this reasoning is applicable to all cases ; indeed it is to be remembered that the sodium salts

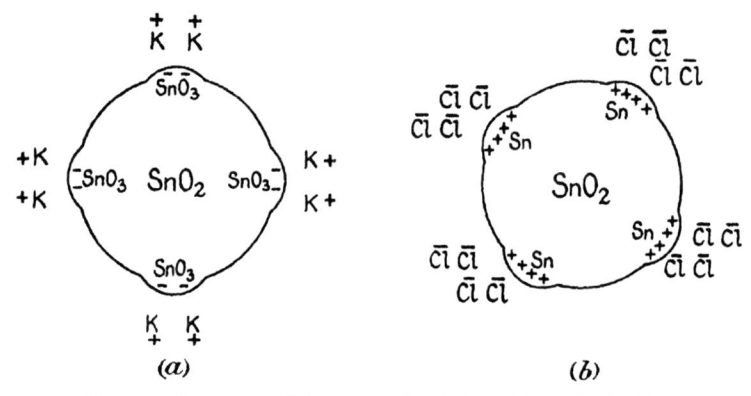

FIG. 4.—Positive and Negative Particles of Stannic Oxide.

forming soluble stannates do coagulate the sols when in higher concentrations, and here it may be supposed that the discharge of the particles is due to the adsorption of the oppositely charged sodium ions.

The behaviour is in line with the theory of Langmuir and Harkins, according to which the molecules of the boundary layer always tend to make the transition of the phases gradual. The stannates turn their stannate ion towards the stannic oxide particle and their potassium ion towards the water.

When a positively charged stannic oxide sol is prepared by peptization with hydrochloric acid the double layer is probably made up of molecules of stannic chloride turning their stannic ions towards the stannic oxide particle and their chloride ions towards the liquid, as represented in Fig. 4 (*b*).

The theory of peptization and coagulation will be taken up

COLLOIDS

again under the discussion of the stability of colloids in Chapter VIII. One more example may be given here to show that the idea of the double layer being formed partly of material already present in the particle has helped us to understand a number of other phenomena.

It has been established by X-ray examination that the particles of many sols are crystalline and the data obtained allow us to work out the crystal structure or space-lattice of the substance. This knowledge of the geometrical linking of the atoms in the crystalline particle gives a closer insight into the relation of the peptizing ion to the atoms or ions of the space-lattice of the particle.

It is a well-known fact that when dilute silver nitrate and potassium bromide or other halide solutions are mixed in exactly chemically equivalent proportions a curdy precipitate is produced which settles at once, but that if either reagent be in excess, part of the silver halide remains as a sol. This fact enables silver solutions to be titrated without an indicator. At the moment we are not concerned with its practical value, however, and the chief point to note is that if silver nitrate is present in excess a positively charged silver bromide sol is obtained, whereas a negatively charged silver bromide sol results when there is excess of potassium bromide. We may regard the positively charged sol, containing excess of silver nitrate, as containing particles of silver bromide, on which are adsorbed silver ions, while the corresponding nitrate ions form the negative part of the electric double layer. Similarly, in the negative sol containing excess of potassium bromide we may suppose that the particles of silver bromide have a coating of adsorbed bromide ions, whilst the corresponding potassium ions form the positive side of the double layer in the liquid.

We can now consider this interpretation from the point of view of the space-lattice of silver bromide. This lattice is represented in cross-section by Fig. 5, where it will be observed that any silver atom not near the surface is surrounded by six equidistant bromine atoms (four in the plane of the paper, one in front and one behind) and each bromine atom is surrounded by six equidistant silver atoms. Actually, we have reason to believe that the units of the lattice are not atoms but ions.

However, the silver ions in the *surface* of the particle represented by the figure are surrounded by only five equidistant bromide ions (three in the plane of the paper, one in front, and one behind) and so at each of these points the lattice lacks one bromide ion to saturate the affinity of the silver ion. Consequently, when the particle is placed in a solution of potassium bromide containing free bromide ions it tends to add a bromide ion to the silver ion in the surface. But the added bromide ion is already balanced by a corresponding potassium ion, which therefore takes up a position opposite the bromide ion. By repetition all over the surface of the particle the result is the formation of an electric double layer consisting of bromide and potassium ions. This

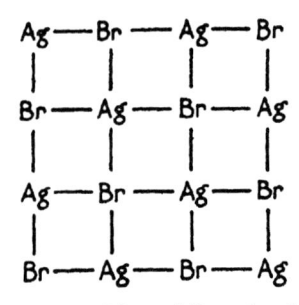

```
Ag—Br —Ag—Br
 |    |    |    |
Br—Ag—Br—Ag
 |    |    |    |
Ag—Br —Ag—Br
 |    |    |    |
Br—Ag—Br—Ag
```

FIG. 5.—Neutral Particle of Silver Bromide

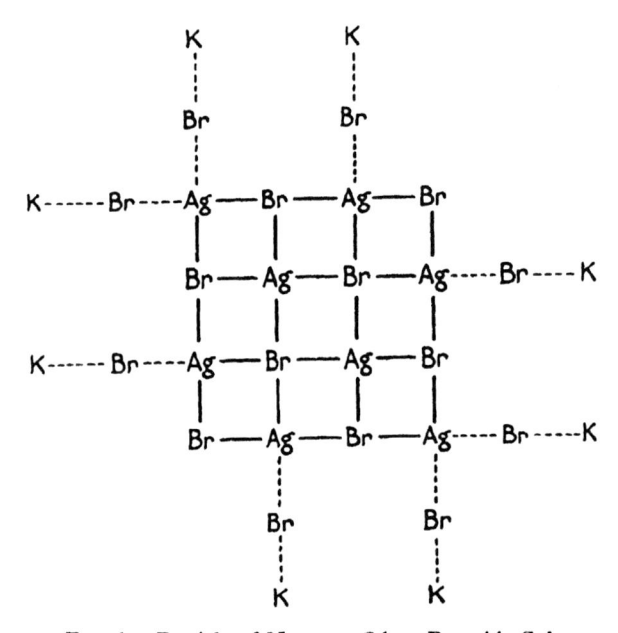

FIG. 6.—Particle of Negative Silver Bromide Sol.

state is represented in Fig. 6, where the dotted lines show the adsorbed ions.

If silver nitrate is in excess, the silver ions are added to the

COLLOIDS 65

surface bromide ions, and nitrate ions form the outer negative sphere, the particle being positively charged. This state is represented in Fig. 7.

It will be observed that in this representation of the facts the forces causing the adsorption are not essentially different from those which preserve the space-lattice. Another point is that the feature of an electric double layer formed in this way is an orientation of the adsorbed ions. It is very probable that not all the adsorbed molecules of the peptizing electrolyte form part

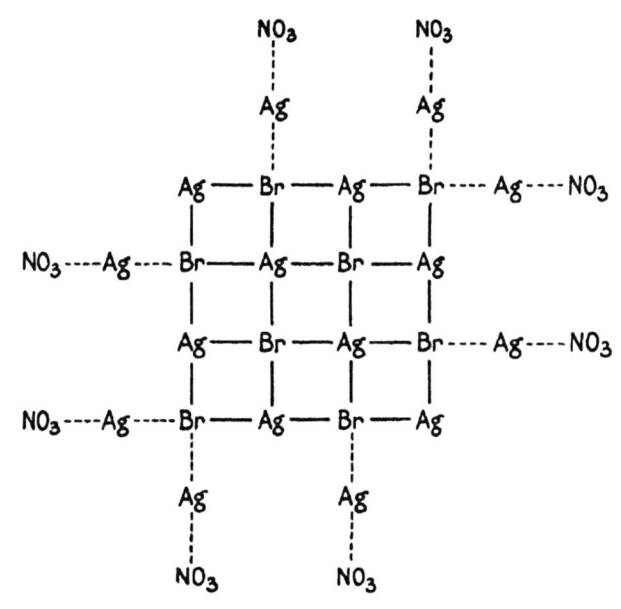

FIG. 7.—Particle of Positive Silver Bromide Sol.

of the double layer but only that part of it which takes up an oriented or polar position.

Although the application of the ionic theory in one form or another to hydrosols has met with such great success, it has its limitations, and this is nowhere more apparent than in the study of organosols in non-dissociating media such as benzene. These systems also contain charged particles, and here perhaps we have to do with electronic rather than ionic disturbances. We do know at least that large differences of potential can be set up between metals and organic liquids such as petrol, and there are cases recorded of serious fires produced by an electric spark

F

passing between two dissimilar metals when pouring petrol from one metal container to another.

The Constitution of Metal Hydrosols. Newer views of the constitution of hydrosols of the noble metals can now be considered. The impression will have been gained from Chapter III that a gold hydrosol consists of minute fragments of pure gold bearing a negative electric charge and floating freely in the dispersion medium, but that certainly does not represent the true picture of colloidal gold.

Fuchs and Pauli (*Kolloidchem. Beih.*, 1925, **21**, 195) have shown that progressive dialysis of a large number of gold sols prepared by different methods is accompanied by the formation of hydrogen ions. The concentration of these was always about 10^{-5} N and it was clearly demonstrated that they were not derived from carbon dioxide in the air or from the substance of the containing vessels or dialysis membranes. It is concluded that the ions are the partners of the gold particles, which are considered to be covered by the anions of a gold acid.

More recently, Eirich and Pauli (*ibid.*, 1930, **30**, 113) have made an extensive experimental study of the conditions governing the stability of gold sols prepared by the electrical disintegration method. In experiments on the preparation of gold sols in very dilute hydrochloric acid solutions it was found that between 3×10^{-4} N- and 10^{-5} N-hydrochloric acid the stability of the sol increases with the concentration of the acid. The addition of small quantities of potassium chloride has a stabilizing effect, chlorine water stabilizes only at a concentration less than $10^{-4} N$, whilst sulphuric acid and gold chloride diminish the stability or flocculate the sol. Similar sols prepared in very dilute sulphuric acid were less stable and coagulated completely in less than twenty-four hours, but the rate of coagulation was retarded by the addition of hydrochloric acid. Sols of gold in dilute potassium sulphate solutions were also prepared, but were very unstable. Sols prepared in the purest conductivity water were exceedingly unstable, but were rendered more stable by the addition of very dilute hydrochloric acid; very dilute sulphuric acid hastened coagulation. Gold sols were also prepared in dilute solutions of chlorine, potassium chloride, bromide, and iodide, but it was not possible to obtain sols in dilute potassium fluoride solutions. Gold sols in potassium nitrate solution could not be prepared unless the concentration of the salt was greater than 10^{-3} N. These results indicate very clearly the importance of the anions which give the negative charge to the gold particles.

The former view, that the production of colloidal gold by reduction of solutions is essentially the crystallization of a supersaturated molecular solution of gold has been called into question by von Weimarn (*Kolloid-Z.*, 1929, **47**, 231). Colloidal solutions of gold iodide may be prepared by adding a very dilute solution of chloroauric acid to a boiling dilute aqueous solution of iodine, or by mixing dilute solutions of chloroauric acid and potassium iodide in the presence of potassium hydroxide. Using potassium thiocyanate in place of potassium iodide, sols of gold thiocyanate are produced. These sols *change spontaneously to gold sols*, thus casting doubt on the usual simple explanation of the production of gold sols. In a review of his work on gold sols, extending over a period of many years, von Weimarn (*Rep. Imp. Ind. Inst. Osaka*, 1929, **9**, 9) shows that his views are opposed to those of Zsigmondy. It is shown that the increase in the mean size of crystals with increase of concentration of the reacting solutions observed in some cases is an indication of the existence of some anomaly in the course of the precipitation process. The colloidal gold is considered to be formed through the intermediary of gold compounds, and the presence of gold compounds in the final sol can be detected by the blue ring given with concentrated ammonia. The question is considered whether it is possible to obtain colloidal solutions of gold without the preliminary formation of colloidal solutions of sparingly soluble gold compounds. Gold is far from being one of the simplest subjects for colloid investigations; it is inclined to the formation of complex compounds, and these are not characterized by stability and they undergo rapid changes in their chemical composition. Von Weimarn emphasizes repeatedly that the importance ascribed by Zsigmondy to the purity of the water used for preparing gold sols is over-estimated; the only effect of increasing purity is to increase the concentration to which the red sols can be evaporated before they become blue. It is concluded that a large proportion of the work on dispersoidal gold synthesis should be repeated, not only quantitatively, but qualitatively, that it is not possible to accept " coagulation values " obtained up to the present time, and that the identification of the synthesis of colloidal gold with spontaneous crystallization is erroneous.

Pauli's work on gold sols has been extended to silver sols (*Kolloid-Z.*, 1926, **39**, 195). It was found quite impossible to prepare a stable sol in pure water by Bredig's method, but that perfectly stable sols resulted in the presence of very dilute potassium hydroxide between the concentrations 1×10^{-5} and 5×10^{-4} N. In the presence of silver hydroxide stable sols were formed only when its concentration was in the neighbourhood of 10^{-5} N. The results show that Bredig's method cannot be regarded as a mere disintegration, but is accompanied by electrolysis, the products of which enter into the constitution of the particle.

A very detailed investigation of the properties of colloidal platinum and the light thereby thrown on its constitution has been made in recent years by Pennycuick (*J. Chem. Soc.*, 1927, 2600; 1928, 551, 2108; 1929, 618, 623; 1930, 1447; *Kolloid-Z.*, 1929, **49**, 407; *Z. physikal. Chem.*, 1930, **148**, 413). The hydrogen-ion concentration of platinum sols, prepared by sparking under conductivity water, was measured by means of the quinhydrone electrode. The p_H values obtained from nine different platinum sols ranged from 4·60 to 4·28, and the measurements showed that the sols become more acid on keeping or on boiling. In the coagulation of platinum sols by a number of acids, flocculation occurred at a nearly constant hydrogen-ion concentration (p_H 3·44–3·69), and evidence has been adduced to show that during this coagulation no acid is removed by the platinum particles, the natural acidity of the sol remaining. These conclusions were confirmed by conductometric measurements. It is suggested that the colloidal platinum particles consist of atoms of platinum together with complex anions of the constitution PtO_4. The corresponding hydrogen ions are considered to be bound as a double layer at the surface, and not free. The natural acidity of the sols is due to a free acid, probably hexahydroxyplatinic acid, $H_2Pt(OH)_6$, which behaves as a strong acid at these low concentrations. The conductometric titration curves of the clear solution after coagulating the platinum sol by freezing is very similar to that of hexahydroxyplatinic acid.

Conductivity measurements show that when salts are added to the platinum sols the cation of the salt partly replaces the hydrogen ion of the hexahydroxyplatinic acid. The acid contained in the surface of the particles is thus responsible both for the negative charge and the stability of the sol. The surface of the particles also contains a more feebly acidic oxide, which is held to be responsible for the protective effect of univalent bases on colloidal platinum and also for the peptizing action of univalent bases on the coagulum. It is considered that in the preparation of platinum sols by Bredig's method the momentarily high temperature of the platinum particles and the electrolysis of the solution are sufficient factors for the formation of stabilizing oxidation products.

CHAPTER VI

PRACTICAL METHODS OF COLLOID INVESTIGATION

There are several special appliances, such as the ultramicroscope, the centrifuge, and apparatus for dialysis and for cataphoresis, which are specially designed for the examination of colloid systems. At the same time a large amount of useful work may be done with the ordinary chemical laboratory apparatus and it is worth while to make a few remarks about this, since special precautions often have to be taken.

In the first place it must be remembered that many colloid systems are extremely sensitive and their stability and general properties may be entirely altered by the presence of small quantities of electrolytes or other impurities. Consequently it is advisable in all experiments and essential in many to ensure that all glass vessels are made of resistance glass. Softer glasses are attacked by water and by other reagents to an extent which is great enough in many cases either to coagulate a sol or to render it more sensitive to coagulation by other reagents. This applies to test-tubes and even to stirring rods.

For the same reason, distilled water should always be used in all preparations or dilutions. In a few instances, such as in the preparation of special gold sols, the presence of minute traces of impurities in the water is of such importance that investigators give special directions for preparation of the water. However, for ordinary purposes, ordinary distilled water suffices. All the vessels should be thoroughly rinsed several times with distilled water immediately before use. Many sols are more or less strongly adsorbed by the glass vessels in which they are stored and then it becomes necessary to resort to more drastic chemical means of cleaning them, such as the use of acid or alkali or oxidizing agent, but the reagent to be used will depend on the particular sol. Traces of gelatin, albumin, etc., left in glass vessels are extremely difficult to remove and will require chromic acid. Gelatin in particular adheres to glass so firmly that in shrinking it may cause the glass to crack. It is also very easy to break beakers in an attempt to remove adherent pieces by means of a glass rod ; consequently, thick walled vessels will often be found serviceable. Silica gel is extremely difficult to remove from glass vessels when once it has become dry. Silicic acid preparations should therefore be kept in wide-mouthed containers and are best removed immediately after use.

Among other points to bear in mind is that the order and rate of mixing of reagents is very important in determining the effect produced. Reference to this will be made in other chapters, but for experimental purposes (and this refers particularly to experiments on coagulation) it is essential to perform all operations in the same order and as far as possible at the same rate. A further point is that the properties of many sols vary considerably with time and all preparations should therefore be dated. It is usual in many investigations to mention the age of the sol.

Great care should be exercised to ensure uniformity of sampling as far as possible. One gelatin leaf may differ quite considerably from another in ash and water content and in other ways.

Dialysis. The essential feature of all dialysers is a membrane which is permeable to electrolytes and impermeable to colloid particles and which is supported in some convenient way. Graham's dialyser consisted of a glass ring or short cylinder open at both ends, over one end of which a parchment membrane could be tied, a groove being cut in the glass to facilitate a tight fit. The sol to be dialysed is put into this receptacle, which is then suspended in a larger vessel containing water (preferably distilled) which is either kept running or is frequently renewed. In any form of apparatus the dialyser must be tested with water for leaks before use, for if the membrane is imperfect or if it is not tightly tied to the support the sol will soon find its way through. Dialysis is in general a rather slow process and is continued until the dialysate gives only a feeble test for the electrolyte that is being removed. As a rule, the process must not be carried too far or the sol becomes very unstable and may flocculate. Only experience can show how far it may be carried in a given instance.

Actually, Graham's dialyser does not provide a very large

surface for dialysis and it is often more convenient to use other types. A common way is to use parchment paper in the form of a tube resembling a sausage skin; these are sold in a flat condition, open at both ends. The skin is bent in the form of a U-tube and placed in a tall beaker. It is then filled with the sol to be dialysed, and since the strain thus imposed on the parchment is liable to spoil it, it is best to fill the beaker with water at about the same rate. The ends of the nearly filled tube may be folded over and suitably clamped.

Parchment is also put on the market in the form of seamless thimbles and this is a most convenient form, especially for dealing with small quantities of liquids. They have the drawback of being rather more expensive. The thimble is filled with the sol and placed in water in a conical flask whose neck is a little wider than the thimble (the thimble swells in water).

Dialysis is far more rapid when membranes of collodion are used and by carefully controlling the conditions of formation of the membrane, septa of widely differing degrees of permeability may be produced. Collodion is a solution of nitro-cellulose in a mixture of ether and alcohol and the usual strength is 3 or 4 per cent. For ordinary purposes, thimbles of collodion for dialysis can be made in the following way.

A thoroughly clean and dry test-tube is filled with collodion, taking care that no air bubbles are formed, and the collodion is then slowly poured back into the bottle by inclining the tube gradually and keeping it rotating constantly. When the last few drops have drained off an even layer should remain adhering to the inside of the tube and this is allowed to dry for some minutes. The amount of drying at this stage largely determines the permeability of the membrane; in general, the collodion should not feel sticky when touched and when the tube is held against a dark background the film should be just visible as a faint blue. At this stage the tube is immersed in water and left for about a quarter of an hour. The film can then be removed from the tube by detaching it at the lip and very gently and slowly pulling it away from the glass wall, the whole operation being conducted under water. To perform this without rupture, especially at the bottom of the tube, requires some care and practice. A useful method of supporting the thimble is to tie it to a glass tube inserted at its upper end.

The seamless thimbles of filter paper made for the Soxhlet extraction apparatus can be made into dialysing thimbles by filling with collodion and, after this has completely penetrated through the walls, inverting the thimble and allowing it to drain, rotating it the whole time. The thimble is allowed to dry for a few minutes and is then immersed in water. These thimbles are stronger than those of collodion alone and are able to stand upright.

If flat membranes are required, they may be produced by pouring the collodion solution on to a surface of mercury.

In the process of *electro-dialysis* the rate of dialysis is increased by causing the electrolyte to travel away under the influence of an electric field. The principle of the apparatus is gained from

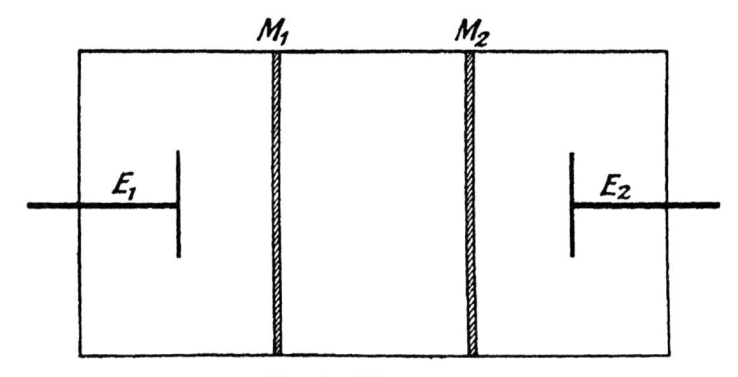

FIG. 8.—Electro-dialysis.

Fig. 8. It consists of three compartments, the middle one containing the colloidal solution. The compartments are separated by two dialysis membranes, M_1 and M_2, and fresh water circulates through the outer compartments in which are inserted two electrodes, E_1 and E_2. Under the influence of the current the ions are carried through the membranes to the electrodes of opposite sign of charge. The purified colloid remains in the central compartment.

Ultrafiltration. The filtration of ordinary coarse suspensions is mainly a sieve action, particles of radius greater than that of the pores being retained by the filter, whilst smaller particles pass through and appear in the filtrate. The pores of ordinary filter papers are too large to arrest colloid particles and consequently colloids pass through the filter in the ordinary way.

However, the filtration of a colloid is not entirely analogous to the action of a sieve because the electric charges on the particles complicate matters. For example, filter paper becomes negatively charged in contact with water and consequently the first small portions of a positive sol which is being filtered will be retained by the filter because the positive charges are neutralized by the negative charges on the filter paper; when an equilibrium has been reached, the rest of the sol passes through unchanged. A negatively charged sol will pass through the filter directly.

The problem of the filtration of colloids is known as *ultrafiltration*, and whilst the effect of the electric charges must always be borne in mind the first thing to do is to obtain an ultrafilter whose pores are much smaller than those of ordinary filter papers. Colloidal gels immediately suggest themselves as ultrafilters, because the pores of these are of colloidal dimensions, and in practice collodion is a very convenient substance to work with.

Ostwald gives a ready and simple means of preparing ultrafilters for ordinary laboratory purposes, which retain most colloidal particles and filter at a satisfactory rate without applied external pressure. The collodion solution recommended consists of 4 grm. of nitrocellulose in 12 c.c. of alcohol and 84 c.c. of ether. An ordinary filter paper is folded in the ordinary way, placed in a funnel, wetted with distilled water and allowed to drain; the filter is then filled with the collodion solution, the excess poured off with constant rotation so that a uniform film of collodion is produced, left to dry for about five minutes and then a second coating of collodion is given in the same way. After allowing a few more minutes to dry, the ultrafilter is ready for use and should give a colourless filtrate with a sol of night blue.

The ordinary rate of ultrafiltration with this arrangement is 1 or 2 c.c. per minute, but the rate may be increased many times by adapting it to a Buchner funnel and applying the suction of the water pump. For this purpose the paper is wetted with distilled water and placed on the perforated plate; it is covered with collodion solution and the excess distributed round the edge of the paper by inclining the funnel and rotating it until the collodion has dried to the extent that it no longer flows. A second coating of collodion is then given in the same way and when this has dried the ultrafilter may be used directly. Neither this nor any other form of ultrafilter must be allowed to dry completely or it becomes useless.

Another method recommended by Ostwald for use with a vacuum pump is as follows. A filter paper, folded in the usual way and placed in a glass funnel, is filled with collodion which is given time to penetrate, the excess being then poured off whilst the filter is slowly rotated so as to produce an even layer. Rotation is continued while the collodion dries; this may be about five minutes, but the rate and amount of drying largely determine the permeability of the ultrafilter. The whole is immersed in distilled water for about a quarter of an hour and the ultrafilter may then be used. The permeability may be altered by placing it in aqueous alcohol solutions of different concentrations.

For ultrafiltration on a larger scale, Bechhold (*Z. physikal. Chem.*, 1907, **60**, 257) has devised a metal apparatus in which the sol is forced through a membrane under a pressure of a few atmospheres. The membrane is made of hard filter paper impregnated either with collodion or with hardened gelatin. In the former case, the filter paper is immersed in collodion solution of a given concentration, removed, allowed to drain in such a way that an even coating is produced, and submerged in water which is constantly changed. It is then ready for use. The gelatin ultrafilters are made by immersing the filter papers in a gelatin sol (2–10 per cent.) at a known constant temperature, removing and draining as before and allowing the adherent gelatin to set to a gel; they are then transferred to a cold 2–4 per cent. solution of formaldehyde for about twenty-four hours.

In either case, the concentration of the sol of collodion or of gelatin is the chief factor determining the permeability of the membrane, and by keeping other conditions constant it is possible to prepare a series of ultrafilters of varying degrees of permeability and consequently to carry out fractional ultrafiltration. The use of this in giving an approximate idea of the size of the particles of sols has already been mentioned (p. 44), but it must be remembered that the results are open to criticism.

The size of the pores of the ultrafilters can be calculated from measurements of the pressure required to force air through the membrane saturated with water or by measuring the volume of water forced through unit area of the membrane in unit time by a known pressure.

The following list of substances is selected from a table given by Bechhold and showing the order in which the disperse phase of the sols will be retained by ultrafilters of decreasing permeability.

Prussian blue	1 per cent. gelatin sols
Platinum sols (Bredig)	Hæmoglobin
Colloidal ferric oxide	Serum albumin
Casein (in milk)	Diphtheria toxin
Colloidal arsenious sulphide	Colloidal silicic acid
Colloidal gold (about 40 $\mu\mu$)	Litmus
" Collargol " (colloidal silver)	Dextrin
Nuclear gold sols	Crystalloids

In considering this table it must be remembered that it refers only to the particular sols used by Bechhold. Sols of Prussian blue, ferric hydroxide, arsenious sulphide, etc., prepared under other conditions may have very different particle sizes and might give quite a different order. It may also be repeated that in addition to the size of the particles adsorption and electric charge play important rôles in ultrafiltration.

The Ultramicroscope. The principle of the scattering of light by colloid particles and the examination of this scattered light in the ultramicroscope has already been described on p. 36 and it is necessary only to give a brief description of the practical arrangement and use of the ultramicroscope. As with all apparatus, more can be learnt in a few minutes' examination and practice with the real instrument than from a great deal of reading.

The observation of the light scattered by the particles against a dark background is ensured in the original slit ultramicroscope of Siedentopf and Zsigmondy by strongly illuminating the particles laterally. The arrangement of the apparatus is shown diagrammatically in Fig. 9. The light from a powerful electric arc A passes through a lens L_1 and falls on to an adjustable fine slit S, of which the lens L_2 and the objective O throw a greatly diminished image in the interior of the trough T which contains the sol. A very small portion of the sol is thus brilliantly illuminated and the vertically scattered light is observed in the microscope M. Since no direct light enters the microscope, the appearance seen in the field is a number of bright discs of light against a dark background.

The sol to be examined is placed in a cuvette, which has a small window in front for the entry of the illuminating beam and another on top facing the microscope for the observation of the scattered light. The provision of a funnel and pinch-cock enables the ready washing out of the cuvette.

This arrangement requires a good deal of apparatus, occupies rather a large space, and powerful lighting is essential, but it is considered by many to be the best form for ultramicroscopic examination. The visibility depends not only on the difference in refractive index of the two phases, but also on the intensity of illumination. Bright sunlight is the best form of illumination, but since this is somewhat capricious it is safer to rely on the

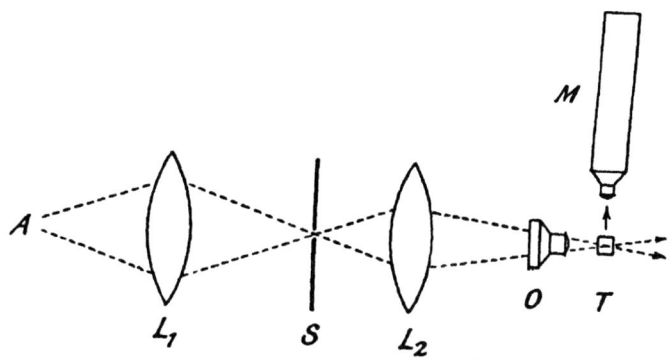

FIG. 9.—Arrangement of Slit-Ultramicroscope.

electric arc light. In bright sunlight, particles of about $5\mu\mu$ diameter can be observed provided that the difference in optical properties of the phases is great.

In the slit ultramicroscope an exceedingly thin *slice* of the sol is illuminated laterally. Just as in ordinary microscopical examination a thin section of the specimen is cut off mechanically, so in the slit-ultramicroscope a thin slice is delimited optically. In a recent modification of the slit-ultramicroscope, the cuvette is eliminated, the objective of the horizontal illuminating system being so close to the objective of the microscope that a drop of the sol can simply be suspended between the two objectives. A greater resolving power (down to $4\mu\mu$) is obtainable with this immersion ultramicroscope, but in the freely suspended drop rapidly changing images are obtained.

Less costly, and, from some points of view, more convenient forms of ultramicroscopes are made by fitting to the ordinary microscope a suitable condenser provided with a central stop which cuts off the rays of light which would normally travel directly into the microscope. There are several types, but the principle will be understood by reference to Fig. 10 which shows the path of light through the *paraboloid* condenser, a simple form. It is well known that parallel rays entering the paraboloid condenser axially are reflected into the focus of the paraboloid, and the position of the upper face of the condenser is so arranged that this focus falls on the surface of the microscope slide, which has to be of a definite thickness.

A central stop at the lower face excludes part of the incident light and allows only such rays to pass as will strike the surface (after reflection) under an angle greater than the critical angle, so that they would be totally reflected by the top of the condenser if in contact with air. By placing cedar oil between this surface and the microscope slide, the light passes through the latter and through the drop of sol placed on it for examination, but is totally reflected at the cover-glass which rests on the drop of liquid. Hence, no light will enter the microscope if the liquid is free from particles and the field will be dark ; the particles of sols will scatter the light, however, and the scattered light will be visible in the microscope.

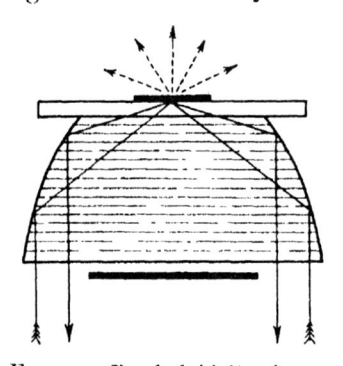

FIG. 10.—Paraboloid Condenser

Cataphoresis. A practical method for the demonstration of the sign of charge of the colloid particles has been described on p. 50. For many purposes, the microscopical or ultramicroscopical method is better when measurements of the cataphoretic migration velocity are required. This can be carried out quite simply by providing an ordinary microscope slide with a pair of parallel electrodes of platinum foil connected by insulated leads to a source of low voltage such as a couple of lead accumulators. Higher voltages are not required, as the distance between the electrodes is not so great as in the macroscopic method. A drop of the sol is placed so as to touch both electrodes and then a cover-glass is put on and the drop is illuminated with a dark-field condenser and is examined through the microscope.

Determinations made by either the macroscopic or the microscopic method are subject to certain errors and these have to be guarded against, particularly in observations in the ultramicroscope, where the movement of individual particles is followed. In the first place, the glass walls of the cell are charged with respect to the liquid and consequently particles near the cell wall may even travel in the reverse direction; this difficulty is obviated by choosing particles near the middle of the liquid. Further, when the particles are near the electrodes they lose their charges and are generally precipitated in its vicinity, but they may also acquire the sign of charge of the electrode and then set off in the reverse direction. On their way, these reversed particles will eventually come into contact with particles having the original sign of charge and then neutralization and mutual precipitation occurs. Another source of danger is that sols generally contain electrolytes and these accumulate at the electrodes during prolonged cataphoresis as in ordinary electrolysis, the catholyte in general becoming richer in hydroxyl ions and the anolyte richer in hydrogen ions. The presence of these ions and the difference in their concentration can make considerable changes in the magnitude of the charge on the particles. These difficulties are avoided by restricting observations to particles roughly equidistant from the electrodes and carrying out the measurements as soon as possible after switching on the current. The best plan is therefore to confine attention to a middle portion of the sol and to avoid unduly prolonged cataphoresis.

Viscosity. Viscosity measurements play a large part in colloid investigations since a great number of sols, particularly of the so-called " natural colloid " type, are characterized by high viscosity. The apparatus most often used is a simple one known as the Ostwald viscosimeter and measures the *relative viscosity* of a liquid. In the case of a sol, the viscosity of the sol is determined in relation to the viscosity of the dispersion medium, generally water.

The Ostwald viscosimeter is illustrated in Fig. 11 and consists of a wide glass tube joined by a bend to a straight glass capillary tube which leads to a bulb holding generally about 2 or 3 c.c. of liquid and provided with an inlet tube, a constriction being placed where this joins the bulb. Marks are etched on the tube at this constriction and at the junction of the bulb and the straight capillary (A and C).

The instrument is used by putting a definite volume of the liquid to be examined into the wide tube by means of a pipette, and sucking it into the narrow tube and bulb by means of a piece of rubber tubing placed at the top of the latter. The liquid is drawn up so that it fills the bulb and rises slightly above the upper etched mark, where it is held by pinching the rubber tube. The hold on the tube is then released and the time of fall of the liquid meniscus *between the upper and lower etched marks* is observed by means of a stop-watch.

The time taken for the liquid to flow through the capillary is proportional to the viscosity provided that the driving pressure is the same. In the present apparatus, where the liquid falls under its own weight, the pressure is proportional to the density of the liquid. The relative viscosities (η_0 and η_1) of two liquids can therefore be obtained from measurements of the times of fall (t_0 and t_1) of the liquids from one mark to the other, and the densities of the liquids (d_0 and d_1), using the formula

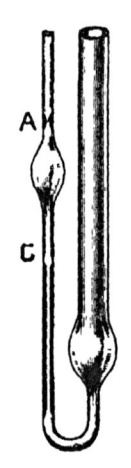

FIG. 11.—Ostwald Viscosimeter (Baird & Tatlock).

$$\eta_1 = \frac{\eta_0 t_1 d_1}{t_0 d_0}.$$

Viscosity varies very greatly with temperature and therefore strict temperature control is essential in these measurements.

The viscosity of many kinds of sols is not so well-defined a physical quantity as that of pure liquids like water, glycerol, or aniline, for in many cases different values are obtained according to the method of measurement and an important observation is that the viscosity varies with the rate of shear of the liquid. This subject is not completely understood yet, and further reference will be made to it in Chapter XII, but it may be mentioned here that various other forms of viscosimeter have been devised which measure this property under different conditions. Some of them are modified capillary viscosimeters of the Ostwald type and others are based on quite different methods.

A well-known type is the Couette viscosimeter, and this is particularly useful in giving a means of determining the value of the viscosity at different rates of flow of the liquid. The liquid to be investigated is contained in a cylinder which is rotated at a known speed by means of a motor, and a metal cylinder is suspended in the liquid from a torsion fibre which bears a mirror. The inner cylinder is turned through a certain angle which can be determined by directing a beam of light on the mirror and measuring the deflection of the reflected beam on a scale. At equilibrium, the torsion in the fibre is balanced by the frictional force, which is proportional to the viscosity and to the angular velocity of the outer cylinder.

CHAPTER VII
CLASSIFICATION OF COLLOIDS

Systems of classification seem to be both the delight and despair of physical science. When new phenomena or new things are discovered an attempt is at once made to classify them or group them with other phenomena or things which are known to have similar characteristics or to behave similarly under certain conditions. Even if the new properties or behaviour cannot be correlated with anything else it is felt that it should fit into some classification and is therefore made a class of itself, or is said to be " anomalous." The reason why hydrogen has been crowded out of the Periodic Table of Elements has worried many a chemist. We impose law, order, and analogy on everything and expect Nature to keep to our impositions ; generally she does not, and we apologetically exclaim, " This law can only be taken as a broad generalization " and " There is no definite boundary, but rather a continuous transition between these groups."

The last sentence, which sums up most classifications, is especially applicable to the classification of colloids. One wonders, Why bother to classify colloids ? In fact, is our tendency to classify a disease or mark of undevelopment of the human mind, having no counterpart in the world of existence ? The answer to these questions is beyond the scope of this book and moreover beyond the scope of physical scientists, but most investigators will answer that, however illogical they may be, systems of classification are worth while and do serve a useful purpose in that they help us to remember things and stimulate us to look for more. Possibly, with greater development of mind, we may be able to get on without pricking the position of our knowledge on squared paper and to survey the chart of Nature as one continuous and consistent whole.

Remembering these limitations, we may consider some of the attempts which have been made to classify colloids. Several

CLASSIFICATION OF COLLOIDS

of them have made no permanent mark on the history of the study of colloids and we shall restrict attention to three which are found generally useful, two of them being universally employed. In the first place, on p. 11 we have considered the division of all possible colloid systems into types characterized by the states of aggregation (solid, liquid, or gas) of the disperse phase and dispersion medium. We are now concerned only with the classification of sols of either liquid or solid particles in a liquid dispersion medium (generally water) on the basis of their general behaviour.

The problem is what general property or behaviour to choose as being pre-eminently characteristic of the colloidal state in order to form a basis for comparison, and since it is difficult to reach agreement on such a matter several classifications exist. Whilst the degree to which a given property is shown differs by small degrees from one colloid to another, it is often found that the properties are inter-related or go in sets, so that entirely different modes of classification may ultimately give approximately the same division. Fortunately, there is this rough correspondence in the classification of colloids.

There are several notable features which differentiate a gelatin sol from a gold sol and although these two examples may almost be taken as extremes of the diversity of colloid properties they may be regarded as prototypes of two groups of colloids. The properties of other colloids resemble to a greater or less extent one or other of these extremes, or, unfortunately for the scheme of classification, may fall about half-way between them. There is a practical advantage that the greater number of known colloids do resemble one or other of the prototypes.

The gelatin sol is highly viscous, on cooling sets to a jelly, and is unaffected by small quantities of electrolytes; the gold sol has practically the same viscosity as water, does not set to a jelly on cooling, and is precipitated by the addition of small quantities of electrolytes. Further, the gelatin obtained by evaporation of the sol redisperses in warm water to form a sol once more; on the other hand, the gold obtained by evaporation of the gold sol does not form a sol again on treatment with water.

Reversible and Irreversible Colloids. This last behaviour was made a basis of classification by Hardy, who divided sols into *reversible* and *irreversible* colloids, according to whether the residue after evaporation redispersed in water giving a sol once more. Gelatin is a reversible colloid and gold hydrosol is an irreversible colloid. Unfortunately, the reversibility of a colloid depends on arbitrary conditions under which the criterion applies. Some undoubtedly irreversible colloids under the above definition are coagulated and redispersed reversibly by other methods, such as the action of electrolytes.

The fact that reversible colloids, like soluble crystalloids, go into solution spontaneously might lead to the supposition that they are closely related, the main difference being one of molecular weight. There is a lot to be said for this point of view, but at best it is only part of the truth. There are many reasons for believing that in many colloidal solutions of this type, such as hæmoglobin, the disperse phase does consist of molecules, which are however so big that they will not pass through parchment membranes and which exhibit the other phenomena associated with colloids. Svedberg's recent ultracentrifugal investigations throw an interesting light on this point. It is certainly not correct, however, to suppose that the particles of disperse phase in the reversible colloids are smaller than those of the irreversible colloids. They may or may not be, depending on the conditions, but highly disperse gold sols in general contain particles which are smaller than those of many reversible colloids such as gelatin. These considerations certainly make a distinction between molecularly and colloidally disperse systems more vague, but in general large molecules appear to have a tendency to group themselves into still larger aggregates. This feature is doubtless a contributory cause to the gel formation which is so well developed in these chemically complex, reversible colloids, and probably plays an important part in the building up of organic tissue.

As a rule, the reversible colloids are not greatly affected by small quantities of electrolytes; they may be precipitated by larger quantities of salts, but even then the effect is generally reversible and when the salt is washed away a sol may often be obtained again. Salts of heavy metals may be much more powerful precipitants and the effects are often specific for certain ions, the precipitation also being irreversible in some instances, so that again the system of classification ceases to be of universal applicability.

On the other hand, the irreversible colloids are usually precipitated by quite small concentrations of most salts, bases, and acids, and this change is typically, though not universally, irreversible. Another point, which is not completely understood at present, is that the addition of a reversible colloid to an irreversible colloid may confer reversible properties on the latter.

Whilst this system of classification is not employed to the extent of the two which follow, it is quite usual to refer to reversible and irreversible colloids, having regard to some particular conditions.

Hydrophilic and Hydrophobic Colloids. Perrin considered the stability of sols as a basis for classification and called them *hydrophilic* and *hydrophobic*, according to whether they " love " or " hate " the dispersion medium—water. Gelatin is a hydrophilic colloid, because it disperses spontaneously in water and shows no great desire to come out again ; gold is a hydrophobic colloid, since it does not disperse spontaneously in water and when a sol has been made the particles are very easily removed from the dispersion medium, e.g. by the addition of a trace of electrolyte. In order to make the terms applicable to dispersion media other than water they were altered by Freundlich to *lyophilic* and *lyophobic*.

It will be evident that the behaviour of sols towards electrolytes is the most important criterion in this classification, and whilst this behaviour is certainly one of the most characteristic of all colloid phenomena, there are certain specific effects which detract from the absolute value of the criterion. For example, some sols, whilst sufficiently stable towards many electrolytes to be ranked as hydrophilic, are very sensitive to the presence of certain specific ions. Consequently, it is very difficult to say whether many sols (silicic acid, for example) are hydrophilic or hydrophobic. Further, if stability is considered in regard to coagulation by electrolytes, the classification is not applicable to sols in dispersion media of non-dissociating solvents. Nevertheless, this form of classification is widely used.

Emulsoid and Suspensoid Colloids. Another classification, quite as much in vogue, is due to Wo. Ostwald, who divides colloids according to whether the particles of the disperse phase are liquid or solid. The two classes are called *emulsoid* and *suspensoid*.

It has already been remarked that one type of colloid (the gelatin type) is characterized by high viscosity of the sol, whilst the other (gold type) has a viscosity differing little from that of the dispersion medium. Actually, the measurement of viscosity is the experimental basis of Ostwald's classification, and it is believed that high viscosity in a disperse system is indicative of liquid disperse particles. There is some evidence for this in coarser systems, for a suspension of finely divided solid matter, such as a precipitate of barium sulphate, does not offer much resistance to stirring, whilst emulsions which consist of two or more immiscible liquids intimately dispersed by shaking are generally very viscous and creamy. An extreme case is given by emulsions of 99 per cent. of mineral oil in 1 per cent. of soap solution, which can actually be cut into cubes.

We know that the particles in gold sols and many other of the suspensoid colloids of low viscosity are solid, because they give the X-ray diffraction spectra of crystalline bodies. It is also reasonable to suppose that the particles of an oil-in-water emulsion are still liquid and that such a system constitutes a different type, or emulsoid colloid. But the question remains, whether the particles of disperse phase of sols of gelatin, agar, albumin, etc., are liquid and consequently give rise to the high viscosity. There is considerable division of opinion about this. In the dry state we know these substances as solids, but on placing in water the colloid swells and imbibes much water in the process; in doing so, it loses much of its rigidity and in general has less " solid " properties. On warming, the substance disperses to a sol, and there is reason to believe that the particles still go on absorbing more water, because the viscosity of the sol increases with time. The final result may easily be liquid droplets dispersed in water, the particles of disperse phase consisting of heavily solvated aggregates of gelatin or other substance. Further evidence is adduced by addition of a concentrated solution of a very soluble salt to a gelatin sol—a process by which water is removed. The effect is to throw down a stringy coagulum of gelatin, which on standing coalesces to a liquid layer. It seems reasonable to suppose that before coagulation, when this phase contained still more water, it was in the form of liquid droplets. However, the liquid-liquid phase theory of this type of sol is not universally accepted, although it has a large body of supporters.

Another difficulty arises in that the difference between liquid and solid properties is not so well marked when exceedingly small particles are considered, for at high dispersions surface tension becomes a predominating factor and the great specific surface of tiny droplets of liquid may confer on them the rigidity associated with solid substances. Similarly, a small particle of solid subject to strong surface forces tends to assume the rounded contour and perhaps some other properties of a liquid surface. This may be the reason why mercury hydrosols, which are emulsoids in this classification, have more nearly the properties of suspensoid sols.

For practical purposes, emulsoids and suspensoids correspond fairly closely with lyophobic and lyophilic colloids and, since neither form of classification is by any means absolute, both methods of nomenclature are frequently used, and both will be used in this book. Although there is continuity from emulsoids to suspensoids or from hydrophilic to hydrophobic colloids, it is convenient to use these classes as the extreme members of each have very different properties.

The following table gives in a brief form the chief differences which characterize the two classes of hydrosols.

Property	Emulsoid.	Suspensoid.
Viscosity	High	Almost the same as the dispersion medium
Concentration possible	High	Always low
Electrolytes	Small Quantities no effect. Salted out by large amounts	Precipitated by small amounts
Coagulation	Set to a jelly	Form coarse granules

Bary (Rev. gén. Colloid., 1930, **8**, 289) has advanced the hypothesis of an "evolution zone" in the dynamic equilibrium of polymerides, an increase in the number of molecules being regarded as "positive evolution" and a decrease as "negative evolution." The theory of molecular evolution is applied to lyophilic colloids, which are regarded as matter in the evolution zone. Lyophilic colloids consist of chain-like polymerides, the ends of the chains being saturated with molecules acquired from the dispersion medium; they are regarded as existing in two states, according to whether their evolution is positive or negative, or whether the number of particles is increasing or decreasing. A marked tendency towards positive evolution is characteristic of a peptizable colloid, the velocity of evolution being modified by temperature, pressure, and the chemical properties of the medium; under certain conditions the direction of evolution may be reversed to negative, leading to coagulation or gelation. The process of positive evolution begins with the swelling of a gel and ends at molecular dispersion. Typically lyophobic colloids dispersed by mechanical or electrical means are not regarded as being in the evolution zone; they neither swell nor form jellies. Colloids prepared by chemical methods, such as the metallic hydroxides and sulphides, are in the zone of evolution, but the direction of their evolution is negative.

CHAPTER VIII
STABILITY OF COLLOIDS

Effect of Gravity. Having concluded from all experimental investigations that a gold hydrosol consists of minute particles of gold of a certain range of diameter suspended in water, the problem arises why these particles of a substance so much heavier than water do not sink rapidly under the influence of gravity, leaving the pure dispersion medium.

The rate of fall of particles through a liquid under the influence of gravity depends on the size of the particle, the difference in density of the particle and the liquid medium, the viscosity of the liquid, and the value of the gravity constant where the observation is made. This is summed up in Stokes' law, which can be expressed

$$V = \frac{2r^2(s - s')g}{9\eta}$$

where V is the velocity of the particle of radius r and specific gravity s, s' is the specific gravity of the liquid and η its viscosity, and g is the gravity constant. It will be seen from the equation that $(s-s')$ may have a positive, negative, or zero value, according to whether the disperse phase is heavier, lighter or of the same density as the dispersion medium, so that the particle may fall, rise, or remain stationary. The equation also shows that the velocity is inversely proportional to the viscosity and is directly proportional to the square of the radius of the particle. When the particle is falling in the vicinity of a wall (and generally it is falling close to a glass wall when observations are being made), a correction has to be added to the equation.

The equation is valid for the particles of suspensions and thus offers a means of determining the diameter of such particles. Applying it to the particles of a gold sol of $10\mu\mu$ diameter and substituting in the equation the known values of specific gravity, viscosity, and gravity constant, the velocity of fall of such particles under the influence of gravity works out to be 0·014 mm. per hour, or about 1 cm. in a month. This is a very slow rate of fall and the result indicates that the mechanical stability of ultramicroscopic dispersions is very considerable, but actually a rate of fall of this order is not observed in sols containing particles of these dimensions. In absence of electrolytes or other causes which produce a coarsening of the particles, colloids are far more stable than is indicated by application of Stokes' law and under favourable conditions may be kept for many years without marked change.

It is obvious that there must be factors which confer on the colloid system a greater stability than accords with its dimensional mechanical properties. These factors are the Brownian movement, which is very highly developed in the smaller ultramicrons and is of the nature of a true diffusion, and the electric double layer associated with each particle, which tends to prevent aggregation of the particles. It is therefore to be expected that a given sol will be more stable the higher the dispersity and the greater the charge on the particles, and in general this is found to be so. Any influences which reduce the charge on the particles cause aggregation to larger units and if these reach a certain size they fall under gravity at a rate consistent with Stokes' law and the colloid is thus precipitated. We have already seen from Odén's work on monodisperse sulphur sols (p. 34) that the more highly disperse sols (those containing smaller particles) withstand to a greater extent the coagulating power of electrolytes.

Effect of Electrolytes. Numerous references have already been made to the fact that sols of the hydrophobic or suspensoid type are readily precipitated by quite small quantities of electrolytes, and from the work discussed in Chapter V it may be taken as firmly established that the presence of electrolytes has a powerful effect on the potential difference between the particle and the dispersion medium and that this is closely connected with the stability of the sol.

The pioneer investigations of the coagulating effect of electrolytes on hydrophobic colloids were carried out by Picton and Linder (*J. Chem. Soc.*, 1895, 67, 63–74 ; 1905, 73, 1906), Schulze (*J. prakt. Chem.*, 1884, 27, 320), and Hardy (*Z. physikal. Chem.*, 1900, 33, 385) and have led to some very important generalizations. From a survey of their experimental work, the following facts stand out and have been confirmed in numerous later researches by a large number of workers.

(1) In the coagulation of a hydrophobic colloid by an electrolyte the active precipitant is the ion bearing a charge of opposite sign to that of the colloid particles. For example, in the coagulation of the negative arsenious sulphide hydrosol by electrolytes the cation has a predominant effect and the nature of the anion has comparatively little effect, whilst the anion is all-important in the coagulation of the positively charged ferric hydroxide sol.

(2) The active precipitant is carried down in part by the precipitate or coagulum and its presence can be detected therein by analytical methods.

(3) A certain minimum concentration of electrolyte is necessary to produce coagulation.

(4) The coagulating power of the active ion increases very greatly with its valency, the minimal electrolyte concentration necessary to produce coagulation being far lower for multivalent ions than for ions of lower valency.

The regularities concerning the sign and valency of the coagulating ion are embodied in what is often called by German authors the "Schulze rule," and by English authors the "Schulze-Hardy rule."

The simplest way to determine the coagulation values of a series of electrolytes for a given sol is to place 10 c.c. of the sol in each of a number of small beakers or test-tubes of resistance glass, and then add to each 10 c.c. of electrolyte solution at different concentrations, which are made up beforehand. The tubes or beakers in which coagulation has occurred after a specified time are noted and then a second series is started with concentrations of electrolyte ranging from that of the last beaker showing no coagulation to that of the first beaker showing coagulation. It is customary to express the value as the number of millimols of electrolyte per litre necessary to cause coagulation in a given time and it is usual to compute the concentration for the total volume of sol plus added electrolyte.

It should be mentioned that it is not an easy matter to define the onset of coagulation, or whether it is more logical to take the first appearance of turbidity or of precipitation of the coarser particles. The terms flocculation, coagulation, and precipitation are generally used rather indiscriminately. The first appearance of turbidity will also depend on the method of observation and particularly on the intensity of the light in which it is viewed. Further, there is an important time effect in coagulation, and a more dilute electrolyte solution is found to precipitate a colloid if given longer time. Besides this, the rate of adding the electrolyte alters its apparent flocculation value; often, the colloid gets "acclimatized" to an electrolyte when the latter is added slowly and more will be required for coagulation than when the electrolyte is added at once. In spite of these difficulties, it is possible to obtain consistent and reproducible coagulation values for an electrolyte with respect to a given sol by performing all operations under as nearly as possible the same conditions. A fixed criterion of coagulation, a fixed time-interval, and the same order and rate of mixing are the essential conditions and give quite satisfactory results, for at best the coagulation values are only relative to a given sol and have little absolute significance.

In the following table, adapted from the experiments of Freundlich (*Z. physikal. Chem.*, 1903, **44**, 129; 1910, **73**, 385), the flocculation values of a number of electrolytes for an arsenious sulphide sol are set forth and illustrate both the validity of the Schulze-Hardy rule as a broad generalization and also some of its defects. As flocculation value is expressed in millimols of electrolyte per litre representing the *minimal* amount required

Electrolyte. Univalent cations.	Flocculation value (Millimols per litre).	Electrolyte. Bivalent cations	Flocculation value (Millimols per litre).
LiCl	58	$MgSO_4$	0.81
NaCl	51	$MgCl_2$	0.72
KCl	50	$CaCl_2$	0.65
KNO_3	50	$SrCl_2$	0.63
$\frac{1}{2}K_2SO_4$	63	$BaCl_2$	0.69
HCl	31	$ZnCl_2$	0.68
$\frac{1}{2}H_2SO_4$	30	$(NO_2)(NO_3)_2$	0.64
Aniline hydrochloride	25		
Strychnine hydrochloride	0.5	Tervalent cations	
Morphine hydrochloride	0.4	$AlCl_3$	0.093
		$Al(NO_3)_3$	0.095
Crystal violet	0.16	$\frac{1}{2}Al_2(SO_4)_3$	0.096
New fuchsin	0.11	$Ce(NO_3)_3$	0.080

for coagulation it is perhaps not out of place to point out that ions of highest flocculating *power* have the lowest flocculation *values*. Actually the effect is made more realistic by taking the reciprocals of the flocculation values, but then the units have no real physical meaning.

A glance at this table shows at once that in the coagulation of this negatively charged sol the anion has very little influence, that the flocculation values of cations of the same valency are approximately equal, and that the flocculating power varies to a surprisingly great degree with the valency of the ion. There is no doubt from these experiments and many others that the Schulze-Hardy rule as a broad generalization is on a firm foundation and has a definite physical meaning, and that where exceptions to the rule occur (and there are many) the explanation must be sought in the operation of factors other than the single influence of the ion of opposite sign of charge to that of the colloid.

From the table of Freundlich's results it appears that the univalent cations have a coagulation value of about 55, the bivalent cations about 0·7, and the tervalent about 0·1. Expressing the flocculation values of the uni-, bi-, and tervalent ions respectively as C_1, C_2, and C_3, the relation between these values may be expressed approximately as

$$C_1 : C_2 : C_3 = 8 \cdot 3^3 : 8 \cdot 3^2 : 8 \cdot 3.$$

A general relation of the kind $C_1 : C_2 : C_3 = K^3 : K^2 : K$ had already been found by Schulze and by Hardy. The numerical value of the constant depends on the particular sol.

However, coagulation as an ionic property is not quite so simple as might appear at first sight, for as a rule these simple relations only occur among the lighter ions, and in many instances ions of the same valency combined with the same anion differ considerably in their coagulating power. Conversely, the anion with which a cation is combined may influence the coagulating power greatly, although, as the table shows, the effect is generally small in the case of light elements of similar type.

A point which calls for special mention is the flocculation value of the hydrogen ion, which is so much lower than that of other univalent cations, indicating that acids are stronger coagulants than the salts of the alkali metals. Another feature is the unusually high coagulating power of organic ions; the univalent organic cations shown in the table not only have much lower flocculation values than those for the simple univalent cations, but also show a great variation among themselves. It is evident that specific effects are concerned in the action of these complex ions.

When a simple cation (e.g. potassium) is combined with a series of organic anions this same specific effect is noticeable, as is apparent in the following table, which gives some flocculation values obtained by Freundlich for some potassium salts in the flocculation of an arsenious sulphide sol.

Salt		Flocculation value (Millimols per litre).
Potassium citrate		240
,, acetate		110
,, formate		86
½ ,, sulphate		66
,, chloride		50

Although the coagulating ion for the negatively charged arsenious sulphide sol is the potassium ion, these results demonstrate very clearly that the effect of the ion of similar charge to the colloid cannot be neglected.

So far the results have referred to a negatively charged sol, but in the flocculation of a positively charged sol similar relations are observed, except that in this case the active precipitant is the anion. This is rendered clear by the following table of results obtained by Freundlich (*Z. physikal Chem.*, 1903, **44**, 129) for the coagulation of a positively charged ferric hydroxide sol.

Univalent anions	Flocculation value (Millimols per litre).	Bivalent anions	Flocculation value (Millimols per litre).
KI	16	K_2SO_4	0·20
KNO_3	12	Tl_2SO_4	0·22
KBr	12	$MgSO_4$	0·22
KCl	9	H_2SO_4	0·5
NaCl	9·2	$K_2Cr_2O_7$	0·19
½$BaCl_2$	9·6		
½$Ba(NO_3)_2$	14		
½$Ba(OH)_2$	0·4		

Again, the superior coagulating power of the bivalent over the univalent ion is in evidence, and there is fair agreement among the members of each group. It will be observed, however, that the hydroxyl ion is anomalous in that its flocculation value corresponds with the value for a bivalent rather than a univalent anion. In this respect it resembles the behaviour of the univalent hydrogen ion among the cations.

Notwithstanding these many departures from simplicity, it is a matter of very great interest that in a large number of cases it is possible to formulate a simple relation between coagulating power and valency and also that the coagulating power should vary so enormously with the valency. In this connexion, Freundlich (*Z. physikal. Chem.*, 1910, **73**, 385) has developed a theory which embraces the facts that the coagulating power of an ion is constant for a given valency, and increases with the valency, and also enables us to understand why the increase is so disproportionate. He believes that all the light inorganic ions are adsorbed by the colloid to the same extent. From cataphoretic measurements it is quite certain that this adsorption which brings about coagulation leads to a reduction of the potential difference between the particle and the medium. Consequently, it follows clearly that univalent ions, if adsorbed in equal amounts, will reduce the charge to equal extents and will consequently have the same coagulation values. The same would hold for a series of ions of other valency, provided that complicating factors such as those suggested above are absent. In comparing the relative effects of ions of different valency it might be supposed at first that bivalent anions should have twice, and tervalent ions three times, the coagulating power of the univalent ion ; but the peculiarities of adsorption affect the matter, for it is found that relatively more substance is adsorbed from a more dilute solution.

Adsorption will be discussed more fully in the next chapter, but the curve $A_1 A_2 A_3$ in Fig. 12 indicates how the amount of substance adsorbed varies with the concentration (C_1, C_2, C_3, etc.) of that substance in solution, and it is clear at once that the amount adsorbed from dilute solutions is *relatively* greater than from more concentrated solutions. This curve holds only for a given constant temperature and is called the *adsorption isotherm*. In the present discussion, therefore, the curve $A_1 A_2 A_3$ may be regarded as the common adsorption isotherm for all inorganic ions, with the exception of ions of the heavy metals.

Supposing that an amount A_1 of a univalent cation is required to be adsorbed in order to lower the charge on the particles to the critical potential at which coagulation sets in, then half that amount

($A_2 = \frac{1}{2} A_1$) of a bivalent ion and one-third that amount ($A_3 = \frac{1}{3} A_1$) of a tervalent ion will require *to be adsorbed*. But the peculiar shape of the adsorption isotherm makes it evident that in order *to be adsorbed* in this ratio the salts do not have *to be present* in this ratio. When the amounts adsorbed (A_1, A_2, A_3 in Fig. 12) are in simple ratios, the corresponding concentrations in solution are very disproportionate. This view offers a feasible explanation of the exceedingly great increase of coagulating power with valency.

The reasoning has involved the assumption that a large number of inorganic ions are adsorbed to the same extent, and so long as

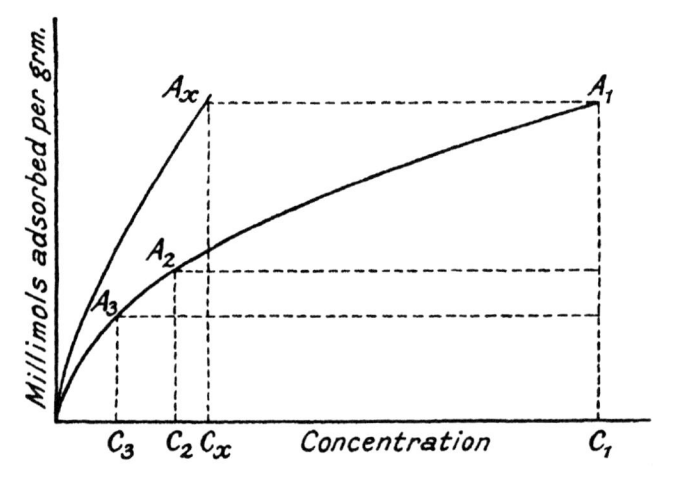

FIG. 12.—Relation between Valency and Adsorption.

this holds it must be expected that the results will be in accordance with the Schulze-Hardy rule. Actually, there are many ions which are known from analytical data to be more strongly adsorbed than others, and it speaks well for the theory that such ions are found to depart from the valency rule. More strongly adsorbed ions must be represented by another adsorption isotherm, which is drawn to the left of the common isotherm in Fig. 12. Its exact position will differ for different ions according to their degree of adsorbability. In this case, an amount A_x is adsorbed at a much smaller concentration C_x; consequently such a strongly adsorbed ion will have greater coagulating power or a lower flocculation value than other ions of the same valency. Actually, hydrogen and hydroxyl ions, organic ions, and ions of the heavy metals are found to be more strongly adsorbed than other ions, and these are just the ions which form the notable exceptions to the Schulze-Hardy rule.

In the table on p. 92, the exceptionally low flocculation value due to the univalent cation of new fuchsin will be noticed. It is found that this ion is adsorbed to an exceptionally strong degree, and actually the coagulation of arsenious sulphide sol by new fuchsin is an excellent method of demonstrating by visual observation that coagulation does really take place by adsorbing the ion of opposite charge, for the precipitate which is formed in this case has not the usual yellow colour of arsenious sulphide, but is distinctly purple, whilst the supernatant liquid when complete flocculation occurs is colourless.

In other less favourable instances the adsorption can be followed by analysis. For example, in some work on the coagulation of arsenious sulphide sol by barium chloride it was found that the precipitate always contained barium but not chlorine. Before coagulation the solution contained 0·1675 grm. of barium; after coagulation it contained only 0·1523 grm.; whilst the amounts of chlorine in the solution respectively before and after coagulation were 0·0865 and 0·0863 grm. The solution was found to be more acid after the coagulation, some of the barium ions having been replaced by hydrogen ions.

It has thus been established experimentally that there is a close relation between adsorption, reduction of the charge on the particles, and coagulation. This enables us to get a picture of what happens during electrolyte coagulation. We must first imagine the colloid particle surrounded by its electric double layer consisting of ions, a certain proportion of which are oriented, so that on the whole the negative ions (in the case of a negative sol) point towards the particle, and the positive ions point outwards towards the dispersion medium.

On addition of an electrolyte, the added cations may be supposed to replace in part the positive ions surrounding the particle, and this process will bring about changes in the electric double layer. The charge on the particle could be reduced either because the new components of the double layer are less highly dissociated or because the degree of orientation is reduced. If either or both of these processes are in operation the successive addition of small amounts of electrolyte will gradually decrease the charge until a critical value is reached, at which flocculation commences. The existence of such a critical value is established experimentally, for it has been found in several researches that in order to produce coagulation it is not in general necessary to reduce the potential of the particles to zero, but to lower it to a certain critical value. On this line of thought, the flocculation value of an electrolyte is the concentration in millimols per litre required to effect such changes in the electric double layer that the potential difference between the particle and the medium is reduced to a certain value. Obviously, this flocculation value will depend on many characteristics of the particular sol under examination, such as the particle size, concentration, and content of stabilizing electrolyte.

The effect of these last factors is well brought out by going a little further with the argument and considering what happens after the potential has been sufficiently reduced. The union of particles, which is the essence of coagulation, is conditioned by two factors—the *probability of collision* and the *probability of adhesion*; in other words, the particles have first to meet and even then the meeting has to be fruitful, or lasting. The probability of collision will depend mainly on the temperature and concentration of the sol, the particles being in rapid motion and approaching each other and retreating continually. The higher the temperature the more lively is this motion, and the more concentrated the sol the less is the distance likely to be traversed by a particle before it meets another. Consequently, it may be expected that rise of temperature and increase of concentration will diminish the stability of sols of the type under discussion.

In these hydrophobic colloids, however, the particles are protected against actual collisions by the electric charge of the double layer, which does not allow the spheres of their electric potential to interpenetrate on account of their equal signs. The probability of adhesion is therefore practically zero, but if there were no charge each collision would result in a union due to the powerful surface forces of such small particles, and the probability of adhesion, or the ratio of the number of unions to the number of collisions, would be unity. By extension, the more the electric charge of the particles is decreased, the smaller is the interference with surface forces and the greater will be the probability of adhesion. In general, this value will not be considerable until the potential is lowered to a certain critical value and then coagulation takes place.

Velocity of Coagulation. The velocity of coagulation has received some study at the hands of several investigators and there is general agreement that after addition of an electrolyte there is first a latent period during which coagulation does not occur, that when coagulation sets in it does so with an initially high velocity, which gradually decreases, and that the velocity increases with the concentration of electrolyte. The kinetics of coagulation have been investigated mathematically by von Smoluchowski (*Physik. Z.*, 1916, **17**, 557, 583 ; *Z. physikal. Chem.*, 1918, **92**, 129) for the case where the probability of adhesion is equal to unity.

A recent development of work on coagulation, due mainly to Kruyt and his pupils at Utrecht, has aimed at studying the kinetics of the " slow coagulation " produced by small concentrations of electrolytes over the course of many hours, rather than the rapid coagulation occurring in a short time interval. It is hoped by these experiments to acquire some more intimate knowledge of the factors affecting the stability of sols. Kruyt points out that rapid coagulation is merely the investigation of a sol that has been robbed of its electric charge and hence can scarcely be called a sol ; he describes the study of rapid coagulation as that of " the anatomy of the dead body of a colloidal solution ! . . . Slow coagulation, on the other hand, is a kind of pathological physiology of the sol."

Effect of Concentration. The effect of the concentration of the sol on the stability towards electrolytes is not at all clear at present. An investigation by Ghosh and Dhar (*J. Physical Chem.*, 1927, **31**, 187) showed that a number of positively charged sols follow the rule that the greater the concentration of the sol the greater is the amount of an electrolyte required for coagulation. Numerous exceptions were found to this rule, depending in many instances on the valency of the precipitating ion. In explanation of the diverse results obtained by these and other authors, Chaudhury (*ibid.*, 1928, **32**, 1231) has pointed out that the dilution of a sol affects its stability in two opposite ways. It is suggested that the diminution of the charge and of the total surface of sol particles with dilution tends to make the sol unstable if the potential at which the sol coagulates and the relative adsorption of all ions on the surface remain unchanged. On the other hand, the greater distances between the particles of a diluted sol favour stability. Recently, Mukherji (*Kolloid-Z.*, 1930, **52**, 63) has made an attempt to find the relation between dilution and stability of a sol by measuring the cataphoretic migration velocity of the particles of a ferric hydroxide sol. The results show that the influence of dilution is very complicated and it is held that the relation cannot yet be discussed successfully.

Adherence. It has already been mentioned that the coagulation of particles will depend not only on the probability of collision, but also on the probability of adhesion. The latter property is very difficult to measure in the case of colloids, but during the last year or two notable advances in the measurement of the adherence of solid surfaces of the same substance have been made by von Buzágh (*Kolloid-Z.*, 1930, **51**, 105, 230 ; **52**, 46). When non-spherical particles of quartz of uniform size settle from a suspension in water or an electrolyte solution on to a plate of quartz which is tilted at an angle to the horizontal, they may readily slide down the plane and accumulate at the bottom of the vessel, or at a lower " angle of tilt " they may rest on the quartz plate. In general, there will be an angle of tilt which is just sufficient to cause the particles to slide off the plate and this is a measure of the adherence of the quartz particles.

There is a close relation between adherence and particle size, and this is connected with a definite attractive force, which is counteracted by gravity and by Brownian movement. Consequently, particles of large size and those of particularly small size fail to adhere to the plate and there is an optimal particle size which, for quartz particles in distilled water, is 3μ. A quartz suspension containing particles of diameter about 12μ was examined with the object of comparing the property of adhesion of the quartz particles to a quartz surface with the electrokinetic potential of the particles. For this purpose the cataphoretic migration velocity of the particles in different electrolyte solutions was also measured. In general, no parallelism of this nature was found, although there was a relation between the discharging effect of small concentrations of the multivalent alkaline-earth, aluminium, and thorium ions and the adherence value. In dilute solutions of alkali hydroxides a relation between adherence number and electrokinetic potential appeared as an equilibrium phenomenon, due to chemical reaction of the alkali at the surface of the quartz. Some experiments with the addition of gelatin showed that the adherence of the quartz particles passes through a minimum with increasing concentration of gelatin. On the other hand, the adherence of glass particles rises with the gelatin concentration, whilst the electrokinetic potential falls. Greater amounts of gelatin have a protective action. It is considered that the electrokinetic potential is not the only factor determining the adherence of the particles, but that the thickness of the double layer and the degree of solvation of the particle are also of importance.

The chief effect of the electrolyte appears to be the alteration of the thickness of the sheath surrounding the particles; in water the layer is thick and easily deformable, whilst in an electrolyte solution it is thin, dense, and not readily deformable. Electrolytes which have a coagulating influence on negatively charged dispersed particles increase the angle of tilt to an amount which increases with the concentration of the electrolyte; electrolytes having a stabilizing influence decrease the angle of tilt. The effect of dissolved chlorides in increasing the angle of tilt is in the order: lithium $<$ sodium $<$ potassium $<$ magnesium $<$ calcium $<$ strontium $<$ barium. For salts of aluminium and thorium, which can reverse the charge on the particles, the curve connecting the sine of the angle of tilt with the concentration of electrolyte passes through a maximum in the isoelectric region and through a minimum in the positive region. In general, there is an antagonism between electrokinetic potential and adherence, but the adherence is controlled both by the electric charge and by the thickness of the layer surrounding the particle. Some preliminary experiments on the effect of adding dehydrating substances, such as ethyl alcohol and tannin, show that these raise the adherence value of the particles.

Acclimatization. The amount of electrolyte required to coagulate a sol depends in many cases on the rate at which it is added. The problem has been studied by Dhar and his collaborators (*Kolloid-Z.*, 1926, **38**, 141) and two kinds of this " acclimatization " have been recognized, for in some cases less electrolyte is required when it is added slowly or in small quantities at a time, and in other cases the amount required is more. The former is termed " positive " and the latter " negative " acclimatization. It seems as though the colloid becomes "used to" the electrolyte when it is presented gradually.

The cause of this phenomenon may not always be the same in different cases, but it is most probably due to changes occurring in the stabilizing electrolytes. Krestinskaja and Jakovlava (*ibid.*, 1928, **44**, 141) have found that when sols of arsenious sulphide are coagulated by the addition of barium chloride solution, more of the electrolyte is required if it is added in small quantities at long intervals. The experimental results suggest that when the barium chloride solution is added gradually the barium ions react with the hydrolysis products of arsenious sulphide, so quickening the hydrolysis. The barium sulphide produced is adsorbed by the colloid, and consequently the barium-ion concentration of the solution is diminished, so raising the critical amount of barium chloride required for coagulation. The explanation given is supported by the fact that hydrochloric acid, which cannot promote the hydrolysis of arsenious sulphide, does not show the acclimatization phenomenon.

In a more recent research, Krestinskaja and Moltschanova (*ibid.*, 1930, **52**, 294) have shown that in all the cases examined by them the phenomenon of acclimatization can be traced to slow chemical changes produced by reaction of the colloid with the added electrolyte. With colloidal ferric hydroxide the added anion is the important factor and it may affect either the external active part of the micelle or the kernel of the particle. In positive acclimatization the added electrolyte forms a soluble salt with the substance of the kernel, with the result that the kernel diminishes and the active part of the micelle increases correspondingly, so that the stability of the sol is increased. This result is brought about by the slow addition of aluminium chloride and nitrate, ferric chloride, and hydrochloric acid. Negative acclimatization is observed in the addition of potassium hydroxide and arsenite, substances which react with the active part of the micelle and increase the kernel, thus decreasing the stability.

Irregular Series. When salts of tervalent or quadrivalent cations are used for coagulating sols of negative charge a peculiar behaviour is often observed. The first small quantities of electrolyte produce no effect until the coagulation value is reached, when precipitation takes place ; the unusual behaviour is that when higher concentrations of electrolyte are used precipitation does not occur, but the charge of the colloid is found to be reversed in sign, and when still higher electrolyte concentrations are used coagulation of this reversed colloid takes place. There is, therefore, first a zone of no precipitation followed by a zone of precipitation, then a second zone of no precipitation with reversal of charge, and finally a second zone of precipitation.

This phenomenon is known as the *irregular series* and is illustrated by the following table, which shows some results obtained by Buxton and Teague (*Z. physikal. Chem.*, 1906, **72**, 57) for the coagulation of a negatively charged platinum hydrosol by ferric chloride.

COAGULATION OF PLATINUM SOL BY FERRIC CHLORIDE.

Millimols FCl, per litre.	Amount of coagulation	Direction of cataphoresis
0 0208	No coagulation	To anode
0 0417	,,	,,
0·0557	,,	,,
0·0833	Complete coagulation	No cataphoresis
0 1633	,, ,,	,,
0 2222	,, ,,	,,
0·3333	No coagulation	To cathode
0·5567	,,	,,
0·8333	,,	,,
1·633	,,	,,
3·333	,,	,,
6·667	,,	,,
16 33	Complete coagulation	No cataphoresis
33·33	,, ,,	,,
83·33	,, ,,	,,
163·3	,, ,,	,,
333 3	,, ,,	,,
666·7	,, ,,	,,

Aluminium chloride produces the same kind of irregular series, and the phenomenon appears when the potential-lowering effect of one ion greatly exceeds the potential-raising effect of the other ion. Since these multivalent ions form salts which are easily hydrolyzed it was once supposed that the effect was due to the oppositely charged hydrolytic colloidal product—ferric or aluminium hydroxide—as described in the next section. But it is found that as a general rule irregular series are produced when one ion is considerably more powerful than the other, either through its higher valency or its greater adsorbability, and organic cations such as those of strychnine nitrate and new fuchsin exhibit the phenomenon in coagulating arsenious sulphide sol in virtue of the much stronger adsorbability of the cation. The same phenomenon is shown in a reversed way in the coagulation of positive sols, such as ferric hydroxide, by sodium phosphate (tervalent anion) and by sodium hydroxide (strongly adsorbable anion).

In the action of ferric chloride on platinum sol represented in the table we can imagine that the negatively charged sol adsorbs the positive ferric ions until the charge on the particles is lowered to the critical potential, when flocculation occurs. In presence of greater amounts of ferric chloride more ferric ions are adsorbed and give a positive charge to the particles so that they are stable but migrate in the reverse direction. At still higher concentrations of electrolyte the adsorption of the negative chlorine ions becomes considerable and lowers the now positively charged particles to the critical potential where the second zone of coagulation occurs.

Some work carried out by Holker (*J. Path. and Bact.*, 1922, **25**, 291, 522 ; *Proc. Roy. Soc.*, 1923, A **102**, 710), which deserves more notice than it appears to have received, would indicate that in the coagulation of some hydrophilic colloids by electrolytes the irregular series is extended into a definitely periodic action. The method of experimentation was to take a series of Jena glass testtubes containing the colloid solution, add to each successive tube an increasing concentration of electrolyte solution and determine by means of a nephelometer the degree of turbidity of the contents of each tube after leaving for four hours. The colloids studied were various animal sera, gelatin, gum acacia, agar, gum mastic, and dialysed hæmoglobin, and the electrolytes used were sodium, potassium, and calcium chlorides, hydrochloric acid, and sodium hydroxide.

The curves in Fig. 13 represent the opacity of diluted human serum in presence of progressively increasing concentrations of sodium chloride and the marked periodicity shown is typical of most of the systems examined. Both the amplitude and frequency increase for the more concentrated solutions of serum.

In the systems mentioned above, the change is not visible to the unaided eye and can be recognized only by means of an instrument for measuring the turbidity, but Holker has one system where the periodic change in turbidity is obvious to the eye. This effect is shown by Wasserman " antigen " when sodium chloride is the added electrolyte; there are three precipitation zones.

Whether the periodic opacity phenomenon of Holker is to be ascribed to the alternate adsorption of ions of opposite sign is not at all clear at present, but it opens up some very interesting

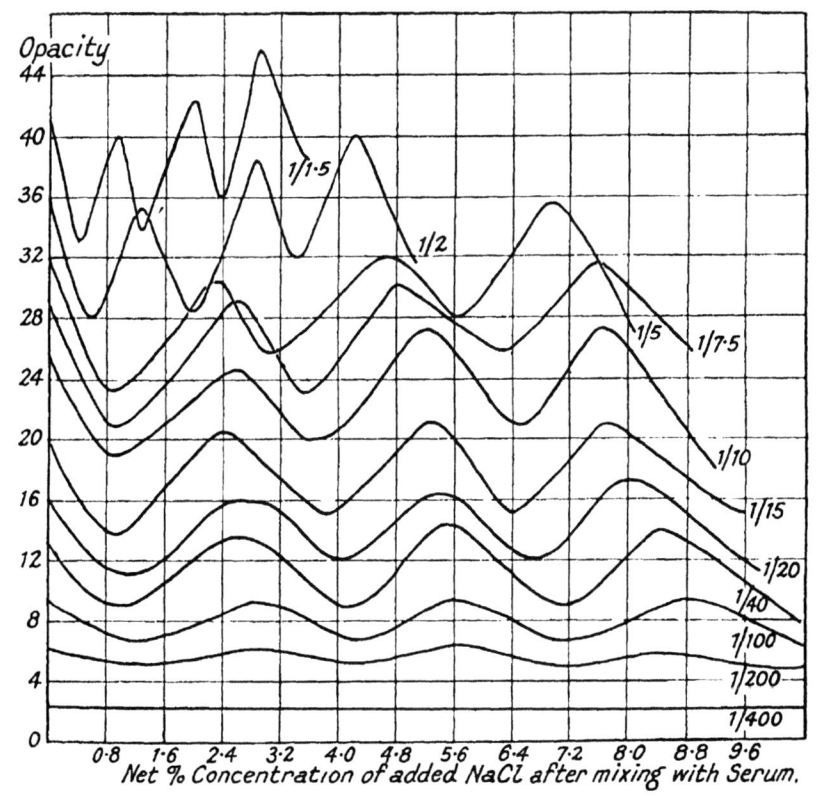

FIG. 13.—The opacity of human serum diluted with progressively increasing concentrations of sodium chloride (*after* Holker)

possibilities and it is a pity that this work has not been repeated and extended by others.

These phenomena demonstrate quite clearly that the stability of colloids may be either decreased or increased by the action of electrolytes; coagulation or peptization may ensue, depending on the conditions. Sen (*J. Physical Chem.*, 1925, **29**, 1533) has shown that a high degree of adsorption and a suitable concentration of the electrolyte are necessary for peptization, and that with the same electrolyte and the same peptizable substance peptization depends to a certain extent on the amount of adsorption. The agglomeration of a precipitate decreases its power of adsorption and its ability to be peptized. Peptization is markedly retarded by the presence of bi- and tervalent negative ions, although the corresponding acids are usually the most highly adsorbed. Univalent acids in general follow the rule that the greater the adsorption, the greater is the peptizing power.

Interaction of two Hydrophobic Colloids. The mutual influence of hydrophobic colloids depends in the first instance on whether the particles of the two sols are of similar charge or of opposite charge. So long as the sols consist of similarly charged particles no important change takes place when they are mixed.

The mixing of two oppositely charged hydrophobic sols is characterized, however, by a definite sequence of events. From the kinetic point of view, the probability of adhesion of the oppositely charged particles on collision will be high and coagulation is therefore to be expected ; but the relative amounts of the two kinds of particles must also be taken into account.

The interaction of two oppositely charged colloids was studied quantitatively by Biltz (*Ber.*, 1904, **37**, 1095) and the conclusions have been confirmed by more recent work. If the particles of one of the sols are in decided excess no flocculation occurs, although the mixed sol may be somewhat turbid, and the resulting sol has the charge, colour, and general properties of the sol which is in excess. The particles of the colloid present in greatest amount appear to cover the particles of the other colloid, for the particles of the mixed sol move as a whole in the electric field in the direction which they would follow if the colloid present in excess were alone. Thus, when a small amount of ferric hydroxide sol is added to a large amount of arsenious sulphide sol the resulting mixed sol is stable and still negatively charged ; similarly, when a small amount of arsenious sulphide sol is added to a large amount of ferric hydroxide sol the mixture is stable and the particles are positively charged.

In the intermediate region, where roughly equal numbers of particles of both colloids are present, partial or complete precipitation occurs. These relations are shown by the following series of results for the interaction of a positive ferric hydroxide sol with a negative arsenious sulphide sol.

MUTUAL PRECIPITATION OF FERRIC HYDROXIDE AND ARSENIOUS SULPHIDE SOLS.

Mg. of Fe_2O_3 in 10 c c.	Mg. of As_2S_3 in 10 c c.	Appearance.	Cataphoresis
0 61	. 20 3	. Turbidity	. To anode
6 08	16 6	. Partial precipitation	,,
9·12	. 14 5	. Complete precipitation	No cataphoresis
15 2	. 10 4	. Partial precipitation	. To cathode
24·3	. 4·14	. Slight turbidity	. ,,
27 4	. 2 07	. No change	. ,,

It is important to note that oppositely charged colloids precipitate each other when they are present in proportions between certain limits and that outside these limits no precipitation occurs. The flocculation is due to mutual discharge of the particles following mutual adsorption. The process is not entirely distinct from the coagulation due to electrolytes, for it is believed that the electric double layer of both kinds of particles is composed of ions. As the particles have opposite charges, their electrical spheres can interpenetrate and the discharge is due to the ionic rearrangements in the double layers, as in electrolyte coagulation.

In practice, the order and rate of mixing the sols are important in determining the result. If the second sol is added in excess very gradually, the optimum of coagulation must necessarily be passed through on the way and then flocculation will result.

Other Coagulating Influences. When the particles move towards an electrode in cataphoresis they flocculate on reaching that electrode, again as a result of neutralization of the charge or reduction to below a certain critical value.

Many hydrophobic sols are coagulated by ultra-violet light, X-rays, or the radiations from radium, but the state of knowledge in this direction is at present very little developed. Experiments seem to show that the coagulating effect of the rays is independent of the sign of the colloid, positively and negatively charged colloids being coagulated at about the same rate.

An idea of the complexity of the action of radiations may be derived from a recent paper by Lal and Ganguly (*J. Indian Chem. Soc.*, 1930, **7**, 513), who have studied the coagulating influence of ultra-violet light. Hydrosols of silver iodide, arsenate, and thiocyanate, gold, silver, vanadium pentoxide, molybdenum-blue, thorium hydroxide, and arsenious sulphide were found to be coagulated by exposure to ultra-violet light, whether the sols were positively or negatively charged. The hydrogen-ion concentration of the ultra-filtrate before and after exposure to the radiation was determined and a detailed examination of the change was made in the case of a silver sol stabilized by tannic acid and a thorium hydroxide sol prepared by dialyzing thorium nitrate solution. In the first case, the coagulation of the silver sol was found to be due to the photochemical decomposition of the tannic acid, which in the presence of a trace of potassium nitrate yields small amounts of ammonia. Similarly, the coagulation of the thorium hydroxide sol and the accompanying change in hydrogen-ion concentration are to be explained by the photochemical decomposition of the stabilizing nitric acid. It is concluded that the coagulation of colloidal solutions by light is due primarily to photochemical decomposition of the stabilizing electrolytes.

Rise of temperature generally decreases the stability of colloids. This is no doubt in many cases due to the dissolution of precipitating electrolytes from the walls of the containing vessel at the higher temperature, but Reid and Burton (*J. Physical Chem.*, 1928, **32**, 425) conclude from an investigation with a copper hydrosol that heat alone is sufficient to cause coagulation. Their results show that there is a definite temperature of coagulation for such a sol and that the coagulation temperature falls with increase in particle size.

Dhar and Prakash (*J. Physical Chem.*, 1930, **34**, 954) have shown that sols of ferric, chromic, aluminium, stannic, and zirconium hydroxides and of vanadium pentoxide and copper ferrocyanide require smaller amounts of electrolyte at 60°, for coagulation that at 30°. A rise in temperature accentuates ageing. It is suggested that this provides an explanation of the greater longevity of cold-blooded over warm-blooded animals.

In some cases it seems that even mechanical agitation can bring about the coagulation of a sol, although the processes involved are not quite clear. Freundlich and Loebmann (*Kolloidchem. Beih.*, 1929, **28**, 391) have summarized some experiments on these lines. Sols of goethite (prepared from iron carbonyl and hydrogen peroxide) and of cupric oxide were found to be coagulated by mechanical agitation, the rate of coagulation increasing with the square of the rate of stirring. The most

COLLOIDS

likely explanation is that coagulation takes place at the surface of the solution, where a certain orderly arrangement of particles may exist; mechanical agitation effects a constant change of surface.

It is very probable that coagulation takes place in two stages, which may be called aggregation and coalescence. If the first stage only is completed the precipitate will in general be capable of peptization, because the particles, which may be supposed to adhere at isolated spots, will readily be separated when they are again charged up by adsorption of a suitable electrolyte or by washing away the coagulating electrolyte. It may be supposed that a process of coalescence of the primary particles follows, the number of points of contact increasing, so that as a rule an old precipitate will not be peptized so readily as a freshly prepared precipitate.

Interaction of Hydrophobic and Hydrophilic Colloids. In general we have to distinguish between two cases : (1) where a relatively small amount of hydrophilic sol is added to a hydrophobic sol ; (2) where the hydrophilic sol is present in excess.

Sensitization. In the first case *sensitization* occurs. This need not involve flocculation, but it is a change in the direction of instability, for the resulting sol is more easily precipitated by electrolytes. For example, if a very small quantity of gelatin sol is added to a red gold sol the colour changes to blue and the resulting sol is much more easily coagulated. Actually, in such an instance these sensitizing conditions are difficult to realize because in any case the concentration of the gold sol is exceedingly small, and for sensitization to occur the concentration of the gelatin has to be something like 1 in 100,000,000. In a similar way, ferric hydroxide sol is sensitized by addition of small amounts of albumin ; the sol retains its stability to all appearances, but the charge on the particles is lowered and the flocculation values of electrolytes in respect of the sol are reduced.

The reason for the sensitization is probably that the sensitizing hydrophilic colloid is itself a colloidal electrolyte, whose ions are adsorbed by the hydrophobic colloid, lowering the potential of the latter to near the critical flocculation potential. In agreement with this is the fact that the sensitizing influence depends largely on the hydrogen-ion concentration of the hydrophilic sol, a factor which determines its dissociation.

According to an investigation by Ghosh and Dhar (*Kolloid-Z.*, 1927, **41**, 229) the sensitization of positively charged colloids, such as ferric, chromium, and aluminium hydroxides and certain colouring matters by the negatively charged colloids, gelatin, albumin, and tannin is due to a neutralization of the charge. On the other hand, the sensitization of the negatively charged colloids, arsenious sulphide, antimony sulphide, silicic acid, silver, and gold by the same agents is due to repression of hydrolysis of the sols through the presence of small amounts of hydrogen ions in the gelatin, albumin, etc.

Zsigmondy (*Z. Elektrochem.*, 1916, **22**, 102; *Z. anorg. Chem.*, 1916, **96**, 265) has studied quantitatively the sensitizing action of certain acid hydrophilic sols on a gold sol by measuring the number of milligrams of the hydrophilic colloid required to produce the colour change from red to blue in 10 c.c. of the gold sol. He calls these values *U-numbers* ("Umschlagzahlen") and a few examples are given in the following table.

U-NUMBERS.

Colloid	U-number
Glycocoll	~80
Histidine	0·1–0·2
Peptone	0·04–0·06
Gelatin	0·002–0·004
Albumose	0·002–0·004
Casein	0·002–0·004

In general, the hydrophilic sols are positively charged in acid solution and negatively charged in alkaline solution. In the determination of U-numbers a positively charged hydrophilic sol is added to a negative gold sol and mutual discharge takes place. This simple explanation is by no means complete, however, for the effect on other negative sols such as mastic may be quite different and it seems that other factors are involved besides electric charge.

Protection. The characteristic effect of the interaction of hydrophilic and hydrophobic colloids is when the hydrophilic colloid is present in excess, and then *protection* occurs. This means that the hydrophobic sol is protected by the hydrophilic sol from coagulating influences; it is rendered more stable to temperature changes and is not readily precipitated by electrolytes. It is a matter of great practical and theoretical importance that a hydrophobic colloid can acquire the stability of a hydrophilic colloid by mixing it with the latter. Another important feature in practice is that these protected sols often have the property that they may be evaporated to dryness, leaving a residue which will redisperse in water to give the sol again; in fact, the formerly irreversible colloid has become a reversible colloid.

Naturally, a phenomenon of such technical possibilities has received a considerable amount of study and it was made the subject of a quantitative investigation by Zsigmondy (*Z. anal. Chem.*, 1901, **40**, 697), who developed a technique for obtaining a measure of the protective effect of various hydrophilic colloids.

The Gold Number. The sharp colour change from red to blue undergone by colloidal gold solutions under the influence of electrolytes makes these sols particularly well adapted for demonstrating and measuring the effect of protective colloids. This colour change is prevented by protective colloids, and so the degree of protection afforded by a given colloid can be measured by determining what Zsigmondy calls the "*gold number.*" The gold number of a protective colloid is the number of milligrams which, when added to 10 c.c. of a standard gold sol, is just insufficient to prevent a colour change on the addition of 1 c.c. of 10 per cent. sodium chloride solution.

Zsigmondy recommends gold hydrosols prepared by the formaldehyde method and having particles between 20 and 30 μ as most suitable for determining the gold number, and gives the following directions for the experiment. It is wise to dilute the colloid until a few tenths of a c.c. will prevent the colour change, but if the protective effect is quite unknown it should be determined roughly before accurate measurements are begun. Then 0·01, 0·1, and 1·0 c.c. of the colloidal solution to be examined are put into three small beakers and thoroughly mixed with 10 c.c. of gold sol. After three minutes, 1 c.c. of 10 per cent. sodium chloride solution is added to each and the contents well mixed. If there is a colour change in the first beaker and not in the remaining two, the gold number must lie between 0·1 and 0·01. For more accurate determinations 0·02, 0·05, and 0·07 c.c. of the protective colloid are taken and the procedure is repeated.

The table given below shows the gold numbers of a number of

protective colloids, and it must be remembered that by definition a low gold number indicates a high protective effect. The reciprocal of the gold number has therefore been put in the third column, since it gives a more realistic impression of the relative protective values of the different colloids. The great variation in gold number from one colloid to another is very striking.

GOLD NUMBERS OF PROTECTIVE COLLOIDS.

Colloid.	Gold number	Reciprocal gold number.
Gelatin and glues	0·005–0·01	200–100
Isinglass	0·01–0·02	100–50
Casein	0·01	100
Gum arabic (good)	0·15–0·25	6·7–4
,, ,, (poor)	0·5–0·4	2·0–0·25
Sodium oleate	0·4–1·0	2·5–1·0
Tragacanth	2 (about)	0·5 (about)
Dextrin	6–12	0·17–0·08
,,	10–20	0·1–0·05
Potato starch	25 (about)	0·04 (about)
Silicic acid	∞	0
Aged stannic acid	∞	0
Slime from the kernel of quince	∞	0

The results are reproducible so long as the same gold sol is employed under the same conditions, but it must be remembered that protective values expressed as gold numbers have reference only to a gold sol and that the order of protective effect of the same colloids towards a different hydrophobic sol, such as arsenious sulphide, would be different.

Nevertheless, the gold number is a quite definite value so long as it is determined under definite conditions and therefore forms an extremely useful characteristic of protective colloids and gives us a means of recognizing impurities in preparations where other methods of analysis are not yet developed and of following the changes taking place in fluids of the animal body. For example, albumins have less protective effect than the globulins; consequently, such changes as occur in the blood plasma in tetanus, where almost all the albumins are transformed into globulins, can easily be detected by determining the gold number.

The investigation of various fractions of the white of a hen's egg gives an illustration of the usefulness of the gold number in characterizing the different colloids present. The original raw material may be separated into different portions by fractional precipitation with ammonium sulphate. The first fraction contains globulin, then follows crystallized albumin and finally amorphous albumins mixed with ovomucoids. Ovomucoid can be isolated by taking advantage of the fact that it is not coagulated when the slightly acid egg solution is boiled. Globulins are insoluble in pure water, but dissolve in dilute salt solutions. The gold number of globulin lies between 0·02 and 0·05, whilst that of the ovomucoids lies between 0·04 and 0·08. Further differences may be detected in the albumin fractions. The first fraction comes down in the form of microscopic crystals, which can be recrystallized and give a gold number of 2–8. The next fraction is amorphous, and not only is it found to be entirely without protective effect on a gold sol, but it actually turns the gold sol turbid and blue without the presence of an electrolyte. A third fraction again appears amorphous, but differs from the second fraction in exercising a high degree of protection on gold sols, the gold number lying between 0·03 and 0 06. Consequently, it has been possible to distinguish by their gold numbers three fractions of albumin which have, so far as can be determined, identical chemical properties.

Theory of Protective Action. Bechhold (*Z. physikal. Chem.*, 1904, **48**, 385) attributed the protective effect to a homogeneous encircling of the suspended particle by the particles of the protective colloid, so forming a protective sheath to each particle of the hydrophobic colloid. According to Zsigmondy, this explanation of the protective effect may be accepted without hesitation so far as coarse suspensions are concerned, but cannot be entertained in respect of ultramicrons of size not far removed from molecular dimensions, such as are encountered in the highly disperse red gold sols. Protection is greater the smaller the particles of hydrophobic colloid, and for a gold particle to be completely surrounded by a larger gelatin particle the latter must have liquid properties, and there are objections to this view.

Billitzer (*Z. physikal. Chem.*, 1905, **51**, 129) concluded that the particles of the two colloids do not unite at all, but the protective colloid adsorbs any precipitating electrolytes. Since it is found that a good protective colloid may counteract a million times its own weight of electrolyte this cannot be regarded as tenable, and in any case it does not explain the changed stability towards other influences such as temperature and evaporation.

Zsigmondy has demonstrated that gold in the form of foil adsorbs on its surface a layer of gelatin which cannot be removed by boiling water and which prevents the amalgamation of the gold with mercury. He therefore suggests as a cause of protection a mutual adsorption of the particles of hydrophobic and hydrophilic colloids. He found (*Z. anal. Chem.*, 1901, **40**, 713) that often the protection is not complete until several minutes after mixing the sols, indicating that a definite time is required for union of the particles. Further, although the concentration of gelatin determines the amount of protection afforded to a gold sol, once the protective action has set in further dilution does not counteract it. The most satisfying experiments were carried out by direct ultramicroscopical examination of the course of protection in the case of gelatin and gold. The gelatin particles of a 1 per cent. solution are so large that they can be seen in the ultramicroscope in spite of the small difference in the optical properties of the two phases. It was found that when the particles of gelatin are large enough the effect is not to form a sheath round each gold particle, but that one particle of gelatin adsorbs several gold particles. Zsigmondy's view of mutual adsorption therefore supposes a union of the particles of the two colloids to form a complex which has the stability of the particles of the protective colloid. The matter can scarcely be considered as clearly understood at the present time.

Preparation of Protected Sols. Protective action is of the greatest importance in colloid chemistry, for by its means many substances can be obtained in a stable colloidal state when the unprotected sol will not keep ; it is also possible to get much higher concentrations of colloids in the protected state and the resulting sols can stand large temperature changes. Another way in which the effect manifests itself is in preventing the growth of primarily precipitated particles, so that many chemical reactions which normally give coarse precipitates may give a colloidally disperse product when the reaction takes place in presence of a protective colloid such as gelatin. In this way it is possible to get stable protected sols of many insoluble sulphates, ferrocyanides, carbonates, etc., which are extremely difficult to obtain in a colloidal form by other methods.

The high concentrations of silver possible in Carey Lea's silver sols (p. 24) are due to the protective effect of the organic substances present. Paal and Amberger (*Ber.*, 1904, **37**, 124; *J. prakt. Chem.*, 1904, **71**, 358; *Ber.*, 1907, **40**, 1392) have developed a method of preparing stable protected sols of many metals and this has received considerable notice on account of the high concentrations of sols it is possible to produce by this means. The protective colloids used are the sodium salts of protalbic and lysalbic acid, two decomposition products of proteins, and the method is specially suited to the preparation of sols of gold, silver, platinum, osmium, palladium, and iridium. Reversible colloids of great stability containing as much as 50 or 70 per cent. of the metal can be obtained and they are not precipitated by even 10 per cent. sodium chloride solution.

As an example, the preparation of colloidal platinum by this process may be discussed. Sodium lysalbate is dissolved in three parts of water and chloroplatinic acid is added and then sufficient sodium hydroxide to dispose of all the chlorine. The reddish-brown resultant liquid is treated with hydrazine hydrate. Nitrogen is evolved, and after standing for five hours the solution is dialysed; it is then evaporated on a water bath and finally in a vacuum. The residue is a black, brittle, and lustrous mass which is soluble in water giving colloidal platinum sol.

Protected colloidal silver preparations, such as "collargol" and "dispargen," are now on the market and have some use in medicine either for intravenous injection or for external application as an ointment in cases of acute rheumatism or pneumonia. Opinion on the efficacy of colloidal silver appears to be divided, but there is some agreement that only the preparations of very fine subdivision are efficacious and that coarser systems are ineffective. In this connexion it is interesting that certain bacilli are retarded in their growth by highly disperse silver sols even at a dilution of 1 in 50,000, whilst coarser sols have no effect. It may be that the effect of colloidal silver is not due to the colloidal properties at all but to the fact that the particles act as a reservoir of silver ions to be had continuously at high dilution. The ions would not be present in high enough concentration to injure the tissues, but would be replaced as fast as used up.

It is evident that not only may the stability of a sol be preserved by addition of a protective colloid but also that a system which would normally consist of coarse particles may be obtained in colloidal form if a protective colloid is present when it is being formed. Consequently, when precipitates are formed by double decomposition in jellies or solutions of gelatin the product will probably be obtained in a protected colloidal state, provided the reacting solutions are not too concentrated. As there is already a colloid present in the form of gelatin, it is not always easy to determine whether the product of reaction is in the colloidal state protected by the gelatin or whether it is present as a highly supersaturated solution, which is in a metastable state but has its apparent stability increased by the presence of the colloid. Whether such a system could be regarded as "molecularly" dispersed is open to question, and our knowledge of the state of supersaturated solutions is so scanty that it is not at present possible to discuss the subject deeply. Ordinary solubility relations may be quite altered in the presence of protective colloids and some systematic work on supersolubility is urgently needed. This matter will be discussed a little further in Chapter XV in dealing with chemical reactions in gels, but an example of the difficulties may be given here.

When equivalent aqueous solutions of potassium chromate and silver nitrate are mixed a red precipitate of silver chromate is produced. When dilute solutions of these reagents in gelatin sols are mixed, no colour change is observed and the resulting solution has the pale yellow colour of the potassium chromate. The red colour is produced if more concentrated solutions are used and the solution obtained is quite transparent so long as a certain concentration is not exceeded; the yellow dilute solutions of silver chromate also turn red when an excess of silver nitrate is added. One view holds that the yellow gelatin sols contain highly disperse colloidal silver chromate protected by the gelatin and that the red sols contain a coarser form of still colloidal silver chromate. According to another view the silver chromate in the yellow solutions is molecularly dissolved and in a high state of supersaturation that could not be maintained in absence of the gelatin, whilst the red solutions contain true colloidal silver chromate. Measurements of the diffusibility and electrical conductivity of the silver chromate in the two forms have given on the whole contradictory results.

An experiment of this type performed by Hedges and Henley (*J. Chem. Soc.*, 1928, 2718) may be mentioned. They found that when solutions of lead nitrate and potassium iodide in agar sols were mixed, at concentrations where the solubility of the resultant lead iodide must be exceeded many times, no precipitation of lead iodide occurred. The excess of lead iodide might therefore be held in colloidal dispersion, protected by the agar. It was also found that when pure, recrystallized lead iodide was warmed with an agar sol and allowed to cool, no solid separated out unless the solubility of lead iodide in water at the ordinary temperature was exceeded about five times. This experiment alone might lead to the view that there is an actual increase in the solubility of lead iodide in water in presence of agar. Considering also the possibility of all sorts of unknown reactions between the "insoluble" salt and the protective colloid, the chemical constitution of which is generally unknown, it is obvious that more work of this type requires to be done.

Ageing. Dhar and Chakravarti (*Kolloid-Z.*, 1927, **42**, 120; *Z. anorg. Chem.*, 1927, **168**, 209) have emphasized the importance of defining the age of a sol in considering its properties. The hydrophobic sols, ferric hydroxide, Prussian blue, copper ferrocyanide, stannic acid, arsenious sulphide, and crystal violet, undergo a fall in viscosity and an increase in conductivity with age. On the other hand, the hydrophilic sols, potassium palmitate and potassium stearate, increase in viscosity with age, whilst the conductivity decreases. Sols of Congo-red and molybdic acid occupy an intermediate position in that there is a very slight fall in viscosity and an increase in conductivity with age. Sols of ferric hydroxide and stannic acid prepared in the cold vary with their age to a much greater extent than do sols of the same substances prepared under boiling conditions; heating hastens the ageing phenomena. Sols of thorium hydroxide, cerium hydroxide, and benzopurpurin manifest decreasing viscosity and increasing conductivity with age. Gelatin sols become more viscous at first, but do not change in conductivity. Both the viscosity and conductivity of silicic acid sols increase with age, this phenomenon being attributed to the aggregation of particles of molecular dispersity and to the release of adsorbed electrolytes. A ceric hydroxide sol, kept until it has stiffened to a gel, ultimately becomes mobile again, the conductivity increasing simultaneously.

Salt solutions which can produce a colloid by hydrolysis also vary with age. The ageing of ferric chloride solutions has been the subject of study by Heymann (*Kolloid-Z.*, 1929, **48**, 25), who has followed the process by making conductivity determinations. The ageing is due to the slow condensation of primary particles of hydrolysis product to form particles of ferric hydroxide of colloidal dimensions, and it was found that the direct addition of ferric hydroxide to the system accelerates considerably the ageing process. Aluminium hydroxide does not exert this effect. The formation of the colloidal particles from the primary hydrolysis product is an autocatalytic process. The ageing of solutions of sodium aluminate is similarly accelerated by the addition of aluminium hydroxide.

A rather different interpretation is suggested by the electrical conductivity measurements of Lesche (*ibid.*, 1930, **52**, 178). The formation of a precipitate during the hydrolysis was not found to affect the course of the ageing curve. The results suggest that the process consists of a gradual transition of the hydrolysis product from a hydrophilic to a hydrophobic colloid, the process being irreversible.

CHAPTER IX
ADSORPTION

In the foregoing pages, numerous references have been made to adsorption without giving any definition of this process. As a matter of fact, a definition is very difficult, and the more one inquires into the mechanism of adsorption the more difficult does it become to formulate a definition which shall differentiate it from other physical or chemical processes. We can at least define the result of adsorption and it is to increase or decrease the concentration of some substance at the interface or surface of separation of two phases. Adsorption is therefore a process intimately connected with surfaces.

It is a matter of common knowledge that many substances, charcoal in particular, can take up large quantities of various gases, such as ammonia, hydrogen sulphide, etc., and this property is used for removing objectionable smells, for obtaining high vacua (the charcoal being cooled by liquid air), and in gas masks in warfare. Obviously, in this process the gas is condensed or its concentration is increased either at the surface of the solid material or in its interior. Not only gases are taken up in this way, but the same substances remove dyestuffs, salts of the heavy metals, acids, and many other substances from solution. Advantage is taken of this property in organic preparations by removing unwanted coloured by-products by means of animal charcoal, and it has also a wide application in industry, for example, in clarifying crude sugar and in removing fusel oil from alcoholic beverages.

As a rule, this process comes to an end in a few minutes and because of its rapidity and in view of several other observations it is believed that the matter taken up is held at the surface of the solid body and does not penetrate through the solid structure. Such a change in concentration at the surface is called *adsorption*, and if there is reason to believe that the substance actually penetrates through the solid wall the term *absorption* is used. *Ab*sorption will obviously be a much slower process and can only follow *ad*sorption. If the taking up of a substance at a surface is unduly prolonged, it is not necessarily due to real diffusion of the substance through the solid, but may be caused by the slow filling up of numerous capillaries in the solid after the external surface has been covered. To include such doubtful cases the noncommittal term *sorption* is sometimes used. The distinction is rendered all the more difficult when the sorbent is in the form of thin lamellæ, for then penetration needs to be only very slight in order to make adsorption a complete absorption, and many colloidal precipitates and gels can be regarded as having such a laminated structure. Nevertheless, we may imagine a more or less rapid surface adsorption, which may or may not be followed by further absorption.

Specific Surface. The enormous importance of adsorption in the study of colloids is due to the large specific surface of colloid systems. By specific surface is understood the ratio of surface to volume, and although this ratio is constant for a given geometrical shape it may not be realized to what a great degree the specific surface of a given amount of material is increased by subdivision. Every time matter is cut, two new surfaces are produced, whilst the volume remains constant, so that the specific surface is increased. The surface of a cube may be expressed as $6L^2$, when L is the length of the cube edge, and its volume is L^3. The specific surface is therefore $6L^2/L^3 = 6/L$. Starting with a cube of 1 centimetre edge, this will have a specific surface of 6 or a total surface of 6 sq. cm. Suppose this cube to be divided into 1,000 cubes each of 1 millimetre edge. The total surface of the same volume is now 1,000 × 6 sq. mm. = 60 sq. cm. If the cube is divided into cubes of colloidal dimensions—say 10^{-6} cm. edge—the total surface of the same amount of solid is easily calculated to be 600 *square metres*—the size of a small garden. It is difficult to realize that the disperse phase of an ordinary sol has such an enormous surface, but once this is grasped it becomes clear that the behaviour of surfaces must play a leading rôle in determining the properties of colloidal solutions. Before discussing this aspect of colloids in particular, therefore, it will be well to outline the phenomena that are characteristic of phase boundaries.

ADSORPTION

The Liquid-vapour Boundary. Certain phenomena connected with a liquid-vapour boundary, such as the appearance of a concave or convex meniscus, the rise or fall of liquids in vertical capillary tubes, and the fact that a relatively heavy small object such as a steel needle will float on water if carefully laid on the surface, are matters of everyday observation. These phenomena were explained fairly satisfactorily as early as 1806 by Laplace, who supposed that the molecules of liquid exert an attraction upon one another so that molecules in the body of the liquid are attracted equally in all directions, being completely surrounded by other molecules, whilst molecules in the surface layer are subject to a one-sided attraction. This gives the liquid an internal pressure which causes it when free to assume the least possible surface, that of a sphere. Van der Waals later modified the theory on the grounds that the transition from liquid to vapour is not discontinuous but continuous. The liquid and vapour do not meet sharply, therefore, in a plane, but some molecules are to be regarded in a stronger attractive position than others so that in the end there is a definite boundary layer of molecules of a certain thickness rather than a boundary surface in the mathematical sense. The internal pressure is a force acting perpendicularly to the bounding layer, but is associated with a *surface tension* which acts in a direction parallel to the bounding surface and which tends to reduce that surface to a minimum. Any increase in surface is against the direction of the force and requires an expenditure of energy, so that a surface is the seat of energy, and if the surface tension be defined as the force operating in the surface, the *surface energy* is the work necessary to increase the surface by unit area.

The methods of measuring surface tension are described in text-books of physics, and the determination is usually carried out by measuring the rise in a capillary, or the weight or volume of a drop of the liquid falling from a tube having a tip of known diameter, or the pressure required to force gas bubbles through a liquid, or the force required to remove a small plate or ring from the surface of the liquid.

One of the most interesting features of the surface tension of a liquid is that the addition of other substances, sometimes in exceedingly minute quantity, generally produces some change in its value. Most often the surface tension is reduced, it may be unchanged, and a comparatively few substances increase the value.

In close relation to this effect of dissolved substances on the surface tension is the fact that the concentration of a solution is not always the same in the interior of the liquid and in the boundary layer. Generally the concentration is greater in the boundary layer, but sometimes it is less. This may be shown by increasing the amount of surface by making the solution into a foam and is exceptionally well shown by some experimental work by Benson (*J. Physical Chem.*, 1903, 7, 532), who led air bubbles through a mixture of amyl alcohol and water, which froths copiously, the foam being carried over to a second vessel and analysed. An excess of about 5·5 per cent. of amyl alcohol was found in the froth. Amyl alcohol is one of the substances which lower the surface tension of water against air. This change in concentration at the surface is what is understood by adsorption, and its relation to surface tension will be described later.

The Liquid-liquid Boundary. That an interfacial tension exists also at a liquid-liquid boundary is evident from the fact that a small amount of liquid freely suspended in another in which it is immiscible forms a spherical drop, indicating a tendency towards minimum surface. Indeed, the conditions at this interface are very closely similar to those at the liquid-gas boundary, and the same methods are in general available for measurement of the interfacial tension. Here also is noted the effect of small amounts of dissolved substances; for example, the interfacial tension between oil and water is decreased considerably by the addition of quite small quantities of soap solution, and more soap is found at the surface of separation than in the bulk of the liquid.

The Solid-gas Boundary. Although this boundary is not yet of much importance in the study of colloids, the study of phenomena occurring in more coarsely disperse systems or with massive material has given considerable insight into the conditions at the surface. In the first place, there are no means at our disposal at present by which a boundary tension at a solid boundary may be measured, but other phenomena which are related to boundary tension are in evidence so that there is every reason to believe that such a boundary tension exists in this case also. For example, it is well known that small droplets of liquids distil to the larger drops and that this is a surface tension effect, for the small drops in consequence of their higher specific surface have more potential energy than the large ones; therefore different amounts of energy will be required to vaporize the small and the large drops and these two kinds cannot exist in equilibrium with the same vapour. The same effect is observed at a solid-gas boundary, for in just the same way small crystals sublime to large ones, so that we conclude that there exists a tension at the boundary of a solid in contact with its vapour.

From kinetic principles Langmuir (*J. Amer. Chem. Soc.*, 1916, **38**, 2221; 1917, **39**, 1885; 1918, **40**, 1361) developed a theory of adsorption in which the adsorbing material is represented as a crystalline space lattice having all its valencies in the interior saturated, whilst at the surface the outer valencies must remain unsatisfied. The essential idea is that these free valencies are responsible for adsorption, and it will readily be seen that the rougher the surface not only is the specific surface for adsorption greater, but the number of free valencies per atom of material increases also. Gas molecules coming within the sphere of attraction of these atomic forces are adsorbed, but they have a kinetic energy of their own which acts in the opposite sense, so that in the end there are two processes going on at the solid-gas boundary—condensation and evaporation—and an equilibrium will be established between the two.

The application of this theory to the catalytic union of gases at solid surfaces is obvious, for the two gases may be adsorbed side by side as atoms and may evaporate joined up as molecules of a new substance. It seems correct to describe the theory as a chemical one, though the whole idea of a distinction between physical and chemical forces becomes vague when we approach molecular dimensions. Normally, we should regard those free valencies at the surface of, say, nickel in their union with oxygen atoms as a chemical linkage; nevertheless, the same valencies presumably hold the nickel atoms together to give the substance its cohesion and other physical properties. If adsorption is to be reduced to a process of chemical affinity it is remarkable that it is not more specific than it is. However, there is supporting evidence on this aspect of the matter from the inert gases helium and neon, which are scarcely adsorbed by charcoal. It should be pointed out also that the inert argon is adsorbed. It is perhaps best at present not to consider in too much detail what is chemical and what is physical.

Another essential feature of the theory is that the adsorbed layer is one molecule deep, for when all the adsorption centres have been covered the formerly free valencies are saturated. In certain instances there is evidence of such unimolecular layers, but it is doubtful whether their existence can be regarded as universal, and there is much experimental evidence which is interpreted to show that adsorbed layers may have a thickness of many molecules. This problem is still receiving widespread examination, but cannot be regarded as settled. It is rarely easy to know exactly the extent of a surface with which one is dealing, and when, as in adsorption experiments as a rule, the substance is finely divided and of irregular shape, the difficulties are correspondingly increased.

The Liquid-solid Boundary. This boundary is of the utmost importance in the consideration of colloid phenomena. Although no means exists for measuring the interfacial tension of this type of boundary, the superior solubility of small crystals over large, causing the large crystals to grow at the expense of the smaller when they are present in a solution saturated with respect to the large crystals, leads us to suppose that such a tension exists.

Adsorption experiments with a pure liquid and a pure solid are difficult to carry out, but the problem has been approached by studying the different volumes taken up by finely divided powders when they are allowed to sediment in different liquids and it appears likely that there is a condensation of the liquid molecules at the surface of the solid. On the other hand, the problem of the change in concentration of a substance dissolved in a liquid in the vicinity of a solid boundary has proved a fruitful study, and is a matter of great technical importance. In what follows, this boundary will be treated in particular, not only because it has been studied more extensively from the point of view of adsorption, but also because adsorption by textile fibres, gels, and sols is generally of this type.

Chemical Phenomena at Boundaries. There appears to be some evidence to show that chemical reactions may be modified in the neighbourhood of a surface. A thorough investigation of this point would be interesting and useful, for there are available artificial methods of increasing the surface in chemical interactions.

It is a familiar fact that when metals are partly immersed in corrosive liquids the degree of corrosion at the liquid-line is often greater than in the body of the solution. The same effect can be observed in the dissolution of substances other than metals. For general dissolution, the protection of the immersed part by the downward flow of the reaction products has been urged as an explanation, and likewise the effect of atmospheric oxygen at the water-line in the corrosion of metals. Doubtless many factors are involved and the reasons may be quite different in different cases, but some experiments by Hedges (*J. Chem. Soc.*, 1926, 831) seem to show that in absence of any of the suggested causes the phenomenon may still be observed.

In certain cases of the liquid-line corrosion of metals the excessive corrosion took place just above the liquid-line, at the place where a film of liquid creeps up the metal and it seemed that the increased velocity of dissolution is a property of this film. Further evidence supporting this suggestion is gained from some observations on the dissolution of magnesium and zinc in dilute acids. At suitable rates of reaction the ascending bubbles of hydrogen coalesce near the end of their course, forming a large bubble on either side of the metal sheet just under the liquid-line. The bubble bursts in a short time, but is immediately formed again. Thus, during the greater part of the reaction there exists just below the liquid-line a thin film of solution in contact with the metal and bounded by a large bubble of hydrogen. Subsequent examination of the metal showed in every case a roughly circular patch indicative of strong preferential corrosion in the region adjacent to the bubble. The conclusion reached from these and other experiments is similar to that formerly suggested by Spring (*Z. physikal. Chem.*, 1889, 4, 658), that the surface of the solvent has an enhanced reactivity.

Another surface difference was observed in experiments on the passivation of iron. It is well known that when iron is immersed in nitric acid above a certain concentration it does not dissolve, but becomes " passive " and inert chemically; it is now pretty generally accepted that this passivity is due to the presence of a very thin film of ferric oxide on the iron, protecting it from attack. Hedges found (*J. Chem. Soc.*, 1928, 970) that in solutions of nitric acid containing 7–14 per cent. of water the production of passivity depends on the rate at which the iron passes through the *surface* of the liquid ; metal which was active as it passed through the surface became passive as it reached the bottom of the vessel. If one piece of iron foil was dropped gently with its flat side to the liquid surface and another piece was thrown edge-ways through the surface, the first piece dissolved with great vigour, whilst the second piece was unattacked. This phenomenon is probably related to the liquid-line corrosion described above, for liquid-line corrosion is very noticeable in this system. Evans (*J. Chem. Soc.*, 1927, 1020) considers that this is to be explained by the protective oxide film tending to leave the metal-liquid interface and collect at the liquid-air interface, and this explanation seems to meet the particular case admirably.

As a further example of the modifying influence of a liquid surface, Hedges has observed that in crystallizing mercuric iodide by cooling a hot saturated solution in potassium iodide, red crystals form on the bottom and sides of the beaker, but the crystals forming at the liquid-air surface are yellow.

Positive and Negative Adsorption. Adsorption has been defined as a change in concentration occurring at the interface of two phases. The change may consist of an increase or a decrease in concentration. If it is an increase, that is, if the substance accumulates at the interface, the adsorption is said to be *positive* ; if the surface concentration decreases and the substance tends to accumulate in the body of the phase, the adsorption is said to be *negative*.

The Gibbs Formula. The relation between the effect of a dissolved substance on the surface tension of a liquid and the change in concentration of the boundary layer was first deduced by Willard Gibbs from thermodynamic principles (*Trans. Conn. Acad.*, 1876, **3**, 391). He arrived at the following expression :

$$U = -\frac{C}{RT}\cdot\frac{d\sigma}{dC}$$

where U is the excess of substance in the surface layer, C is its concentration in the bulk of liquid, σ the surface tension, R the gas constant, and T the absolute temperature.

ADSORPTION

If an increase in concentration goes with increase of surface tension, then $\frac{d\sigma}{dC}$ is positive and combined with the minus sign on this side of the equation shows that the excess U will be negative—in other words the substance is negatively adsorbed. When increasing concentration reduces the surface tension $\frac{d\sigma}{dC}$ is negative and U becomes positive, indicating that positive adsorption will occur. This is most often the case. The view to which the reasoning leads, and which is supported by all the experimental evidence obtained, is that a substance is positively adsorbed when it reduces the interfacial tension, and negatively adsorbed when it increases the interfacial tension. Regarded from a slightly different angle, a surface is the seat of a surface energy which always tends to a minimum; if a substance can by its accumulation at that surface decrease the energy, it will do so; on the other hand, if it increases the energy, it will be less concentrated in the surface than in the bulk.

The equation also shows that a substance can decrease the interfacial energy a great deal, but can only raise it by a small amount, for the more the substance lowers the surface tension the more it is adsorbed, whereas a substance which raises the surface tension can only exist at the interface in small concentration because of negative adsorption. This is in accordance with experience and the relatively enormous lowering of the surface tension of water by minute traces of surface active substances is a well-known fact. It also follows from the equation that the amount adsorbed decreases as the temperature rises.

Attempts have been made to verify this formula quantitatively by experiment, but the experimental difficulties involved are large. The liquid-liquid interface has proved most susceptible to measurement from this point of view because it is possible to divide one liquid into fine droplets of known dimensions in another liquid, and the results obtained are in fair agreement with the formula. It must be remembered that in applying the theory to solid surfaces there is no means of measuring σ, the surface tension, and the coefficient $\frac{d\sigma}{dC}$ becomes a problem unassailable by experiment and where analogy is the chief support.

As a surface phenomenon, it is obvious that among other

things the amount adsorbed will be proportional to the active surface. As it is generally impossible to measure this, difficulties immediately arise in determining whether the results of experiments with different systems are due to a different extent of surface, to specific differences in the adsorbents or in the adsorbed matter.

The Adsorption Isotherm. The process of adsorption does not go on indefinitely ; it soon reaches an equilibrium and there is a true equilibrium between the amount adsorbed and the concentration of the solute, which is the same from whatever side the equilibrium state is approached. The equilibrium is a true one in the sense that the amount adsorbed per gram of adsorbent is independent of the relative amount of the adsorbent. Some recent investigations, which have led to the formulation of what is called the " solid-phase rule " and to which brief reference will be made later, appear to show that this statement is not always true. As a rule the equilibrium is reached quite rapidly, particularly when agitation of the liquid is employed, and a slow, steady increase in the amount adsorbed may indicate a further process of absorption or even a slow chemical reaction. The adsorbent cannot always be regarded as inert in the ordinary chemical sense and it often contains gases, such as oxygen, which might unite with the adsorbed substance. Adsorbents are often not chemically well defined ; wood charcoal, which is a very favourite one, contains a certain amount of ash and its properties depend in no small degree on this variable factor.

Freundlich (*Z. physikal. Chem.*, 1907, **57**, 385) deduced the following relation between the amount adsorbed and the concentration of the solute :

$$\frac{x}{m} = kC^{1/n}$$

where x is the amount adsorbed, m the weight of adsorbent, C the concentration, and k and n are constants. The ratio x/m is a measure of the amount adsorbed per unit weight of adsorbent, or the relative adsorption. The values of the constants k and n depend on the particular system, but the value $1/n$ has little specific character, falling mostly between 0·2 and 0·7 for a very large range of systems. This is seen from the following table of values of the exponent obtained by Freundlich.

Adsorbent.	Solvent.	Solute.	$1/n$.
Blood charcoal	Water	Formic acid	0·451
,, ,,	,,	Acetic acid	0·425
,, ,,	,,	Benzoic acid	0·338
,, ,,	,,	Picric acid	0·240
,, ,,	,,	Chlorine	0·297
,, ,,	,,	Bromine	0·340
,, ,,	Benzene	Benzoic acid	0·416
,, ,,	,,	Picric acid	0·302
,, ,,	Water	Patent Blue	0·190
Wool	,,	,, ,,	0·159
Silk	,,	,, ,,	0·163

The equation of Freundlich given above refers to a definite constant temperature and is called the *adsorption* isotherm, or sometimes the *concentration function*. The relation between the relative adsorption and concentration is well illustrated by the two following sets of figures, which are typical experimental values obtained for the adsorption of acetic acid in water and benzoic acid in benzene, respectively, by charcoal at $25°$.

ADSORPTION OF ACETIC ACID IN WATER BY CHARCOAL.

Concentration.	x/m	Concentration.	x/m.
0·018	0·467	0·268	1·55
0·031	0·624	0·471	2·04
0·0616	0·831	0·882	2·48
0·126	1·11	2 79	..76

ADSORPTION OF BENZOIC ACID IN BENZENE BY CHARCOAL.

Concentration	x/m.	Concentration	x/m.
0 0062	0·437	0 053	1·04
0·025	0 78	0 118	1·44

These results are plotted in the form of curves in Fig. 14 and this form of curve is characteristic of the adsorption isotherm. The chief feature of the curves is the initial sharp rise, showing that the amount of adsorption is proportionately greatest in dilute solution. This is a matter of the utmost importance, because it explains why excessively small quantities of certain substances can affect the properties of a system to so large an extent. It is probable that this feature plays an important rôle in catalysis, the poisoning of catalysts, the inhibition of chemical reactions, and also in many biological processes. This behaviour is also responsible for the extreme difficulty of removing the last traces of adsorbed matter from a surface by washing, because at great dilutions the equilibrium is all in favour of the solid surface retaining the material.

When the logarithm of the concentration is plotted against the logarithm of x/m, using the values tabulated above, straight lines are obtained, and this is the most convenient way of testing the validity of the adsorption isotherm for any given system. Altogether a great number of systems have been examined from this point of view and in many the simple adsorption isotherm is found to hold, whilst in others it does not accurately fit the

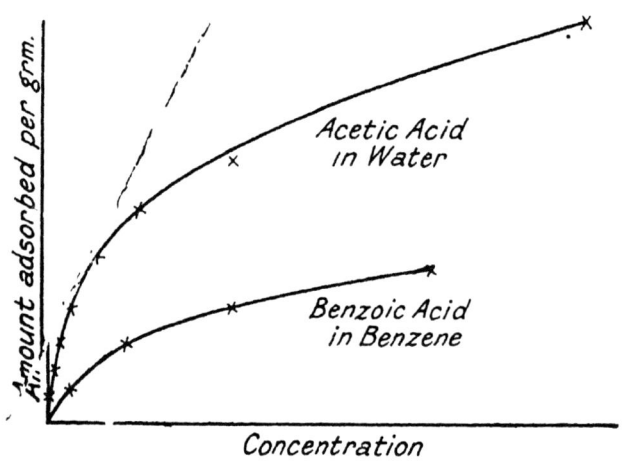

Fig 14.—Adsorption Isotherms

experimental data. Several empirical equations have been devised to meet special cases.

The adsorption equilibrium varies with the solvent employed and therefore it is often possible to remove a dye, for instance, from water through adsorption by charcoal and then reclaim the dye from the charcoal by washing with alcohol. It happens as a rule that solutes are adsorbed to a greater extent from aqueous solution than from other solvents, and this is probably connected with the high surface tension of water. The lowering of the surface tension of water by solutes being greater, the adsorption is also greater. This is illustrated by the following table showing the amount of adsorption of benzoic acid from solution in various liquids and comparing the values with the surface tension of the liquid.

EFFECT OF SOLVENT ON ADSORPTION OF BENZOIC ACID.

Solvent.	Degree of Adsorption	Surface Tension.
Water	4.27	73 dynes/cm.
Benzene	0.55	28.8 ,,
Ether	0.30	23 ,,
Acetone	0.30	16.5 ,,

The simple experiment described below shows this up strikingly and also illustrates the difficulty of washing off the adsorbed material. A few c.c. of a dilute solution of the dye Night-blue are put in a small beaker and swilled round so as to cover a fair amount of the surface. The solution is then poured away and the beaker washed out repeatedly with water as many times as desired. The wash water is not coloured and the beaker appears to be perfectly clean. A few c.c. of alcohol are then poured into the beaker and the liquid will be seen to turn blue immediately, due to the desorption of the dye from the glass walls. This is because the equilibrium for the alcohol-glass surface is less in favour of the solid surface than in the case of water. Of course, a new and reduced equilibrium amount is still left on the glass, even after washing with alcohol.

The tenacity with which this residual adsorbed substance is retained may have far-reaching effects in delicate researches such as investigations of catalysis. A reaction vessel in which a catalytic process has been studied may be quite unfit for another experiment and such vessels may even appear to develop a sort of "personality." To give an example of this effect, Hedges and Myers (*J. Chem. Soc.*, 1924, **125**, 621) found that the addition of minute quantities of chloroplatinic acid to hydrochloric acid in which metals were dissolving with evolution of hydrogen caused a relatively enormous increase in the reaction velocity. Even a concentration of chloroplatinic acid so small as 10^{-9} caused an increase in velocity of reaction of the system Al + N-HCl of more than 20 per cent. In an attempt to investigate this type of catalysis, some measurements were taken with very small quantities of the chlorides of platinum and gold, but further investigation of this interesting case had to be suspended because the reaction vessel could never be used more than once. After it had once contained the platinum salt, subsequent experiments in that vessel always went with a higher velocity than they would have done in an entirely new vessel. This was doubtless due to the adsorption of the platinum compound—or more probably colloidal platinum produced by reduction—on the walls of the vessel. This catalytic activity acquired by the vessel was not impaired by steaming for twenty-four hours.

The degree of adsorption of a solute from a given solvent depends partly on the polar properties of the molecules of the solvent. The relation to be expected between the dielectric properties of solvents and the adsorption of solvents therefrom has been discussed theoretically by Heymann and Boye (*Z. physikal. Chem.*, 1930, **150**, 219), who have also made an experimental investigation. Purified, almost ash-free wood charcoal was used as the adsorbent and it was established that, in the adsorption of benzoic, acetic, formic, and butyric acids, and iodine (violet solutions only) from a number of solvents, weak adsorption occurs in strongly polar solvents and strong adsorption in weakly polar solvents. This antibatic relation between dipole moment and adsorption is not universal, however, and is probably influenced by the mutual affinities of the solute and solvent. This suggestion is supported by the observation that iodine is much more weakly adsorbed from brown solutions in alcohol, ether, and benzene, where it is solvated, than from violet solutions, where there is practically no solvation. A definite and constant relation between dielectric properties and adsorption occurs only in solvents which are closely related chemically, such as a homologous series of alcohols.

Although adsorption is to a certain extent specific, the nature of the adsorbed substance is generally more important than the adsorbent, for it is often found possible to arrange a series of solutes in a certain order of adsorbability, which is the same, perhaps with slight differences, for a large number of adsorbents. This fact enables some general rules to be given as a rough guide to adsorbability. Adsorption is generally slight for inorganic substances with the exception of the halogens, but the salts of the heavy metals are usually adsorbed more strongly than those of the lighter metals. Acids and alkalis are also somewhat more strongly adsorbed than salts. Adsorption is stronger for aliphatic organic compounds and is especially strong for aromatic substances. As a general statement it may be said that increasing molecular complexity goes with greater adsorbability.

As a rule, surface tension decreases with rising temperature and the degree of adsorption is less at higher temperatures. In the cases examined experimentally the temperature effect is in qualitative but not quantitative agreement with the formula of Gibbs. There are some instances of a decrease of boundary tension with rise of temperature, for it is well known that there exist systems of partially miscible liquids exhibiting a lower critical temperature.

Complications ensue when more than one solute is present and this has not been studied in much detail for non-electrolytes, but as a rule both substances are adsorbed to a less degree than if they were present alone. Generally a selective adsorption occurs to a certain extent. From this point of view an electrolyte may be regarded as containing two solutes in the form of the ions, for one of these may be adsorbed preferentially. This gives rise to some important phenomena, as will be shown later. If an adsorbent is placed successively in two solutions, the first of which contains a less strongly adsorbable solute than the second, the second solute may displace the first from the adsorption layer and take its place. As an example, if a dilute solution of the dye fuchsin is shaken with charcoal it is rapidly adsorbed and the liquid appears colourless. If the liquid is poured away and replaced by a dilute solution of saponin, the latter is adsorbed and displaces the dye from the adsorption layer so that the liquid becomes coloured again. This process is sometimes called *exchange adsorption* and is often observable in the adsorption of electrolytes, where it produces some results which it would be difficult to explain on any other grounds.

As a further illustration of the difficulty of testing the validity of the simple adsorption isotherm for a given case it may be pointed out that the amount adsorbed is calculated from analyses of the liquid before and after adsorption—usually by titration. This involves the assumption that there is no change in volume and also that the solute alone is adsorbed and not the solvent. These assumptions are incorrect, but need make little difference so long as dilute solutions are investigated. In more concentrated solutions the errors become considerable and lead to peculiar results. If a series of benzene-acetic acid mixtures of various compositions is examined, at low concentrations the acetic acid is found to be positively adsorbed, but at high concentrations of acetic acid the benzene is adsorbed more than acetic acid, and the amount of acetic acid adsorbed therefore appears to be negative; at some intermediate concentration, equal amounts of benzene and acetic acid must be adsorbed and the composition of the liquid will not change, so that no adsorption will appear to take place.

Lyosorption. The adsorption of the solvent by the adsorbent or of the dispersion medium by the disperse phase is termed lyosorption. This is a process which has received comparatively little study so far, but is one of considerable interest because the solvation of the particles contributes something towards their stability, and in the case of lyophilic colloids is probably the ruling factor.

An investigation by Wo. Ostwald and Haller (*Kolloidchem. Beih.*, 1929, **29**, 354) has shown that when powders are shaken with liquids in which they are insoluble and are then allowed to settle, the volume of the sediment varies with different liquids. The variation in sediment volume is caused by the binding of a layer of liquid by the surface of the powder. The sedimentation volumes of talc, fuller's earth, alumina, silica, magnesium oxide, calcium carbonate, ferric oxide, graphite, and norit stand in the same order for all the liquids tested, being greatest for carbon tetrachloride and decreasing in the order hexane, benzene, chloroform, ether, acetic acid, alcohol, acetone. The volume can be reduced by mechanical pressure or by centrifuging, and is smaller at higher temperatures. By centrifuging the sediment a constant final volume is attained, which, although smaller than the sedimentation volume, varies in the same order throughout the series of liquids. Lyosorption values and the dielectric constants of the liquids investigated were found to vary antibatically, and it is inferred that the force involved in lyosorption is of electronic nature.

The sedimentation volume of quartz particles of microscopic dimensions has been the subject of a recent investigation by von Buzágh (*Kolloidchem. Beih.*, 1930, **32**, 114). The sedimentation volume was measured in water, electrolyte solutions, lyophilic colloidal solutions, and some organic liquids. In pure water the sedimentation volume is independent of the particle size, but in the presence of coagulating electrolytes the sedimentation volume is greater the smaller is the particle size. The sedimentation volume in electrolytes increases with the adherence of the particles and this apparent contradiction is explained by supposing that when the adherence is small the particles can glide over each other and become packed in a small space, whilst when the adherence is great the particles adhere at corners and edges, thus giving an open structure and larger volume to the sediment. It follows also that the value of the sedimentation volume alone cannot in such cases give a true indication of the size of the lyospheres surrounding the quartz particles. An indication of the size of the lyospheres can be obtained, however, when the size of the particle is small in comparison with the thickness of the solvate sheath or when the particles are in the form of oriented platelets, giving a uniform method of packing. Experiments were therefore conducted with plate-like particles of glass and it was established that the sedimentation volume and therefore the thickness of the solvate sheath decreases in the presence of a coagulating electrolyte. The sedimentation volumes of quartz particles in a number of organic liquids increased with the adherence values. In medium concentrations of gelatin solutions the sedimentation volume of quartz particles varied with time, being at first greater than the value obtained in water and later falling gradually to a low value. This behaviour is due to the slow formation of a gelatin sheath round the particles, a thin layer having a coagulating influence and a thicker layer a stabilizing effect.

Electro-adsorption. The previous discussion has been concerned mainly with non-electrolytes and no attempt has been made to consider the influence of electric charges on the adsorption. But, as colloids are typically electrically charged, it is not feasible to leave this aspect out of consideration. The electric charge is probably the most important factor in the following phenomena. A wall of silica acquires a negative charge in water. If a positively charged sol of ferric hydroxide is poured down a column of ignited quartz sand the filtrate is colourless. That the process is not one of ultra-filtration is obvious from the fact that the sand can retain only a certain amount of the ferric hydroxide, and after that the sol runs through unchanged. The same series of events is noticed using a solution of the dye Night-blue, which also is positively charged.

The same type of phenomenon occurs when strips of filter paper are partly immersed in colloidal solutions. Filter paper becomes negatively charged in contact with water, and if partly immersed in a positively charged sol the disperse phase is coagulated on the paper and the liquid rising in the strip is the pure dispersion medium. Negative sols will rise in the paper. For the same reason, positive sols suffer a decrease in concentration when they are filtered through filter paper. These phenomena are obviously closely connected with the mutual discharge of oppositely charged colloids following mutual adsorption as described on p. 106.

Adsorption of Electrolytes. Modern views on the nature of the electrical double layer of the colloid particle and the stability of colloid systems in general make the study of the adsorption of electrolytes by colloids a matter of prime importance. The majority of investigations on adsorption have been carried out, however, with organic substances or substances which are not dissociated, largely because on the whole the adsorption of electrolytes is slight and accurate experiments are all the more difficult to conduct and to interpret.

It was pointed out on p. 95 that Freundlich explained the Schulze-Hardy rule on the assumption that the adsorption of all ordinary inorganic ions is the same. The theory was critically examined by Wo. Ostwald (*Kolloid-Z.*, 1920, **26**, 28, 69), who concluded that whilst it was tenable in a fair proportion of cases, in many instances it does not account for the effects of even closely related electrolytes. On the whole it seems that there is not the great variation in the degree of adsorption of electrolytes that is encountered in other substances, but nevertheless the differences are sufficiently marked to be measurable in favourable instances. Experiments by Odén on the adsorption of nitrates of the alkali metals on charcoal showed that the alkali ions are adsorbed in the following order: $Li < Na < K < Rb < NH_4 < Cs$, suggesting a relation with atomic weight. An increase of adsorbability with atomic weight was also found in the series of halides. Potassium salts gave the following order of adsorption of anions:

$$OH > CNS > ClO_3 > NO_3 > CrO_4 > Fe(CN)_6'''' > SO_4.$$

The fact that electrolyte adsorption is controlled by some influences of a different nature from those operating in the adsorption of non-electrolytes is shown by the behaviour of an adsorbent towards a mixture of electrolyte and non-electrolyte. As a rule, the adsorption of one class of substance is not modified by the presence of the other; the adsorption of an electrolyte is scarcely affected by the presence of a non-electrolyte, and vice versa. Usually, two electrolytes will displace each other when present simultaneously in the same way as was noted for two non-electrolytes, but there are anomalous instances where the two are more strongly adsorbed than when each is present alone.

The experiments showing that ions are adsorbed to different extents indicate that most electrolytes will not be adsorbed as a whole, but that one ion will be more strongly adsorbed than the other. Some important conclusions may be reached from this idea, for suppose an aggregate of uncharged particles adsorbs an electrolyte in such a way that the anion is adsorbed preferentially, leaving an excess of the cation in the solution; the excess cations will arrange themselves as an outer atmosphere and the formation constitutes an electric double layer. In other words, the substance will be peptized, the aggregate separating into unit particles on account of the similar sign of charge taken up, and a negatively charged sol will be formed.

When this idea of the preferential adsorption of certain ions is applied to a colloid system it must be remembered that the colloid particle cannot be considered as at an adsorption zero to begin with; its very stability depends on the layer of adsorbed electrolyte which it already possesses. Consequently, adsorption of a new electrolyte consists in an ionic interchange with the electrolyte already adsorbed, and in some cases it has been shown that electro-equivalent quantities of ions are adsorbed and given off.

Exchange adsorption of this type can cause very considerable changes in the composition of the solution; it is not merely a case of the concentration of solute in the solution being diminished, but also the appearance of a new ion in the solution to replace the ion which has been preferentially adsorbed. A hydrolytic type of adsorption is quite common, hydrogen ions or hydroxyl ions appearing in the solution when a neutral salt solution is exposed to an adsorbent. An example of this is observed with powdered manganese dioxide prepared from potassium permanganate and sulphuric acid and washed until acid can no longer be detected in the wash water. A neutral solution of potassium chloride kept in contact with this manganese dioxide develops an acid reaction, suggesting that the potassium chloride has been hydrolysed. On the exchange adsorption view, what happens is this: manganese dioxide prepared in this way adsorbs sulphuric acid and this is not completely eliminated by the washing. The sulphuric acid is adsorbed in such a way that the sulphate ions are oriented so as to face the particle of solid, whilst the hydrogen ions point outwards towards the solution. On presentation of potassium chloride an interchange occurs, the potassium ions taking the place of the adsorbed hydrogen ions, whilst the hydrogen ions enter the solution and give the observed acid reaction.

If in such an exchange process the substance formed in the boundary layer is feebly dissociated or oriented the electrical double layer will be destroyed wholly or in part, the charge on the particles will be reduced to below the critical potential and flocculation will ensue. This discussion suffices to show the significance of electrolyte adsorption in peptization and in coagulation.

One instance of a practical advantage of exchange adsorption may be given in the realm of agricultural practice. Through a process of exchange adsorption the addition of lime to a soil liberates the adsorbed potassium and renders it available for assimilation by the plant.

The Langmuir-Harkins Theory. Langmuir (*J. Amer. Chem. Soc.*, 1917, **39**, 1848) and Harkins (*ibid.*, 1917, **39**, 354, 541), working independently and from quite different methods of reasoning, arrived at a theory affecting adsorption and other phenomena which may shortly be stated in the following way: Molecules tend to become oriented at an interface.

We may consider Langmuir's point of view first. A drop of oil placed on a surface of water spreads, and when it has spread to its utmost it is accepted (Rayleigh, *Phil. Mag.*, 1899, **48**, 331) that the oil film has unimolecular thickness. Langmuir devised an exceedingly sensitive balance apparatus with the aid of which it was possible to measure the force required to compress this thin film. A known weight of the oil was placed on water in a trough so that the total water surface was not covered after spreading. By moving a strip along the surface of the water the effective area of the trough could be diminished and the force required to effect the decrease of surface could be measured directly by placing weights in a balance pan.

Later and more precise measurements have been made by Adam (*Proc. Roy. Soc.*, 1921, **A99**, 336 ; 1922, **101**, 452 ; *J. Phys. Chem.*, 1925, **29**, 87) and the general result is that the force of compression is connected with the area of the film in a way illustrated by the curve in Fig. 15. Starting with a large area to the right of the figure no force is required to keep the movable strip in place. At *S* it becomes necessary to apply a rapidly

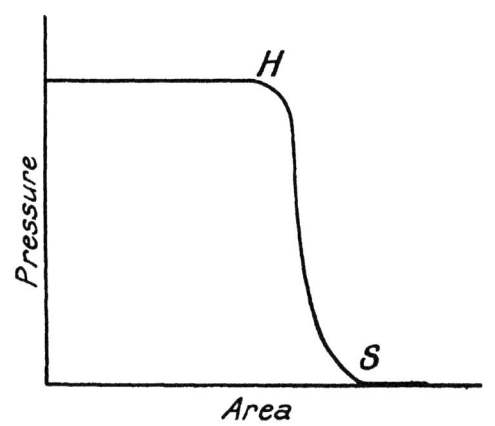

FIG 15 — Relation between Compression and Area of Oil Film on Water.

increasing force for quite a slight decrease in area. Beyond the point *H* no further increase in force is required to reduce the surface of the oil.

The interpretat on of the curve follows best from Fig. 16, where *a* represents the unimolecular oil layer before any compressive force is exerted on it, *b* represents the position at point *S* in Fig. 15 where the molecules practically touch each other and there is great resistance to further compression, and *c* represents the state beyond *H*, where any attempt to decrease the area further results in a crinkling of the film, some of the molecules being pushed up into another layer, just as would happen with a single layer of marbles.

From the form of the curve may be derived a confirmation

of the existence of unimolecular layers, and since the total weight of substance in the film is known and also the molecular weight, the number of molecules per unit area is readily calculated and hence the cross-sectional area of the molecule. The length of the molecule in a direction perpendicular to the water surface can be obtained from the quotient of volume and area. For example, these constants may be determined for oleic acid by dissolving a known weight of oleic acid in benzene, placing a known small volume of the solution on a water surface, and after the benzene has evaporated measuring the surface covered by the unimolecular oleic acid layer.

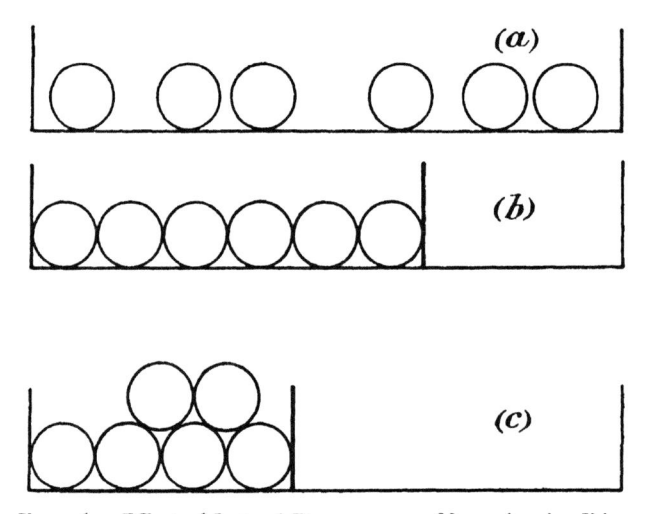

FIG 16.—Effect of Lateral Pressure on a Unimolecular Film.

It is found that the dimensions of the oleic acid molecule in a direction parallel to the surface and in a direction perpendicular to it differ very considerably, and this can only be explained by supposing that the molecules of the unimolecular layer are oriented. If they were arranged haphazardly pointing in all directions, on the average the length of molecule pointing in any one direction would be the same. The measurements given in the following table show, however, that the length in a direction perpendicular to the surface is several times the diameter of a molecule in a direction parallel to the surface. The table shows, moreover, that in a series of long chain carbon compounds the measured length of molecule perpendicular to the surface does

ADSORPTION

increase with the number of carbon atoms in the molecule, but the other dimension remains almost constant.

Substance.	No. of carbon atoms.	Diameter of cross section in μμ.	Length in μμ.
Palmitic acid	C_{16}	0·46	2·4
Stearic acid	C_{18}	0·47	2·5
Cerotic acid	C_{26}	0 50	3 1
Myricyl alcohol	C_{30}	0·52	4·1

These results are readily explained by supposing that the long chains are oriented with their long axes perpendicular to the water surface. The acids and alcohols mentioned are to be regarded as having their carboxyl or hydroxyl groups anchored

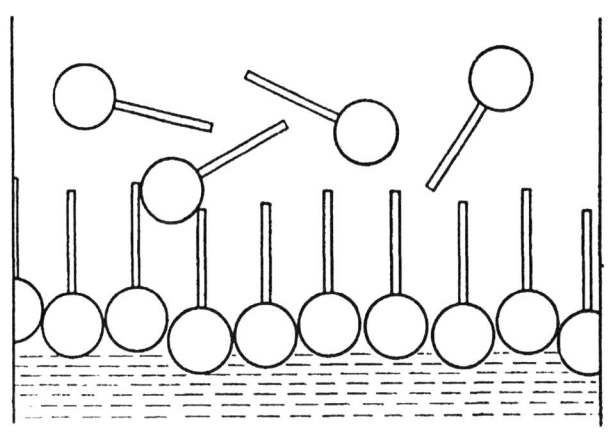

FIG. 17.—Orientation of Molecules at Surface.

in the water surface and their carbon chains pointing away from the water as in Fig. 17. Only substances containing such polar groups are found to exhibit this phenomenon; saturated hydrocarbons, for example, do not spread to a unimolecular layer, but form a lens on the water surface.

Harkins considered the matter from an entirely different point of view. Suppose a column of water of 1 cm.² cross section is pulled in two, then 2 cm.² of new surface are formed, and as the surface tension of water is 73 dynes the work necessary to pull the water apart is 2 × 73 ergs, so, 146 ergs are liberated on reunion, or we may say that the work of *cohesion* of water is 73 ergs per cm.². Similarly the work of cohesion of benzene is 28 ergs per cm.². A benzene-water interface has an interfacial tension of 35 dynes and the amount of work required to break this interface is 73 + 28 − 35 = 65 ergs per cm.², which may be called the work of *adhesion*.

The following table gives data for the work of cohesion and the work of adhesion to water for the aliphatic hydrocarbons and their simplest derivatives. The values are approximate, for they do not refer to a particular substance but to a type of compound. Thus all the higher aliphatic hydrocarbons give a cohesion value round about 45, alcohols about 55, etc.

	Work of Cohesion.	Work of Adhesion to Water.
Hydrocarbons	45	44
Monochlorides	47	81
Alcohols	55	92
Monocarboxylic acids	57	95

When a column of hydrocarbon is broken no orientation occurs because there is no polar group and the work of cohesion is low. A high value is due to the extra work required to overcome the attraction of polar groups which leads to orientation. When the liquid contains a polar group, as in the alcohols and carboxylic acids, orientation of molecules does occur at the place of rupture, but the molecules of the boundary layer arrange themselves with their polar ends turning towards the body of liquid and their non-polar hydrocarbon chains pointing outwards, so that the separation is really between two non-polar parts of molecules, which do not exert a great attraction on each other. Consequently the work of cohesion is not much greater for these than for the completely non-polar hydrocarbon.

Similarly, the separation of hydrocarbons from water takes place without orientation and the work of adhesion to water is therefore not very different from the work of cohesion. The matter is quite different for polar liquids in contact with water, for the work of adhesion is much greater than the work of cohesion. It is supposed that this is due to the extra energy required to overcome the orientation of the molecules at the interface, the polar ends of the organic molecules being anchored in the surface of the water. The polar groups include ionogenic groups such as carboxyl and in addition groups such as − OH, = O, − NH₂, − SH, etc. The polar character of double bonds is shown by the fact that the work of adhesion to water

ADSORPTION

is considerably higher for the unsaturated hydrocarbons than for the corresponding members of the paraffin series. The benzene molecule apparently lies flat on the surface of the water.

The significance of the theory of the orientation of molecules at boundaries is particularly great in the study of adsorption. To take an example, suppose a layer of benzene is poured on water and that butyric acid is added. This dissolves readily in benzene but to a less extent in water. Its carboxyl group is attracted to the water, however, in virtue of its polar nature, and the result is that the molecules tend to collect at the liquid-liquid boundary, where their carboxyl ends can remain in the water leaving the hydrocarbon chains in the organic solvent. In other words, positive adsorption of butyric acid has occurred at the interface.

In the light of the theory we gain a clearer insight into wider matters than adsorption. It is immediately clear why the work of adhesion is the same for all the higher alcohols, for the alcohol side of the interface consists of — OH groups in each case and the pull is against these groups rather than against the molecules. The theory also bears some relation to solubility, for if a strongly polar group is attached to a short hydrocarbon chain as in methyl and ethyl alcohol, acetic acid, etc., it is attracted so strongly that the hydrocarbon chain is itself dragged down into the water and the liquids are completely miscible. Higher alcohols are only partially miscible with water and when the hydrocarbon chain becomes long only the polar end " dissolves " in the water.

This may offer an explanation of why some substances are soluble in a mixture of solvents, though insoluble in any of the solvents taken separately. Part of the molecule may " dissolve " in one solvent and the remainder in another, so that when both solvents are present simultaneously and are miscible the substance will dissolve as a whole.

CHAPTER X

PROPERTIES OF HYDROPHOBIC COLLOIDS

Various properties of typically hydrophobic colloids such as sols of the noble metals have been described in preceding chapters and the conception gained is that of a number of exceedingly small particles of a substance dispersed in a liquid medium, each particle being in violent Brownian movement and possessing an electric charge due to an oriented layer of adsorbed ions, and losing its freedom and stability when robbed of its charge by any process. The size of the particle is fixed between certain limits and in certain instances its shape may be roughly determined from the optical properties of the sol. The fundamental basis of the classification is one of stability and that aspect of this type of colloid has been discussed in some detail.

The impression will have been gained that hydrophobic colloids are less stable in general than the hydrophilic sols and considerable emphasis has been laid on the electric charge in relation to stability. Whilst these views meet with very wide acceptance it must be remembered that the study of colloids is still a very young science and is growing at so rapid a rate that our interpretation of the facts cannot always keep pace with the rate of accumulation of facts. It therefore seems necessary to point out that it is desirable from time to time to review the general ideas of colloid behaviour in the light of increasing knowledge. The explanation of the behaviour and stability of hydrophobic sols in terms of electric charge and adsorption of ions does not stand in danger of overthrow. There is no doubt that the particles are charged, there is no doubt that ions are adsorbed, but the connexion between these two facts is not quite as clear as could be wished and the problem just how far colloid stability is related to electric charge cannot be regarded as settled.

The importance of electrical properties is firmly established, but at least one other essentially important factor in determining

PROPERTIES OF HYDROPHOBIC COLLOIDS

stability is the degree of solvation of the dispersed substance, or in the case of hydrosols the degree of hydration of the particles. That hydration does play an important part seems obvious as soon as we begin to consider the relative stabilities of different types of colloids. Among the very sensitive sols are those of the noble metals, and so far as we know these substances do not combine with water or associate with it in any way. At the other end of the scale we can regard colloids like gelatin and agar as possessing relatively very great stability, and these substances are typically hygroscopic ; they normally contain water, they rapidly take up water from moist air, and when placed in liquid water they imbibe a large amount and increase in volume considerably. Since the massive substance is strongly hydrated when put into water it must be supposed that the dispersed particles are at least as heavily hydrated, and consideration of a large number of cases shows that the more readily hydrated substances form the hydrophilic type of colloid. In fact, the name itself indicates this. To take an intermediate case, metal hydroxides such as ferric hydroxide are usually precipitated in a highly hydrated condition ; in fact, after air-drying a high percentage of water still remains in association with the material. Sols of these hydroxides have a far greater stability than gold sols and in some respects they resemble the hydrophilic as much as the hydrophobic colloids. Silicic acid is another example of a colloid intermediate in properties between the two types and for an inorganic substance it is highly hydrated.

These considerations merely show that it is not wise to put so much faith in the electrical properties of colloids as a guide to their behaviour as to be blind to other facts of possibly equal importance ; they act as an antidote to the essentially electrical treatment which has been followed in the preceding chapters. So far as hydrophobic sols are concerned, we know very little indeed at present about the hydration of the particles, if it exists at all, but the impression is that the molecules of water in immediate contiguity to the surface of the particle are not in the same state as the bulk of molecules in the dispersion medium.

The optical properties of hydrophobic sols have been discussed in Chapter IV and in the present short chapter it only remains to describe some of the remaining physical properties of these systems in so far as they are of a general nature.

L

Surface Tension. The surface tension of hydrophobic sols is not essentially different from that of the dispersion medium. Since quite small quantities of surface active materials have a powerful effect on the surface tension of liquids, the low concentration in which hydrophobic sols are generally obtained cannot be held to have any special significance in this matter.

Viscosity. The viscosity of a system varies with concentration in such a way generally that the low concentration in which hydrophobic colloids are usually obtained is a serious drawback to deducing any definite conclusions from the measurements. As a rule the viscosity of these sols does not differ appreciably from that of the dispersion medium, but in those cases where higher concentrations can be obtained the viscosity is somewhat higher. If the measured differences are to be sufficiently greater than the experimental error, either a sol of exceptionally high concentration must be taken or examination must be made of systems of much coarser dispersion than a colloid.

Among the sols of unusually high concentration may be mentioned Odén's sulphur sols, and the measurements given in the table show the relative viscosities obtained for different concentrations of two sols containing particles respectively of 100 $\mu\mu$ and 10 $\mu\mu$.

Concentration per cent.	Relative Viscosity. 100 $\mu\mu$	10 $\mu\mu$
5	1 20	1 30
10	1 50	1·72
15	2 00	2 38
20	2 75	3·63

When these values are plotted, the curves show that the viscosity increases at a greater rate than the concentration and it also appears that the sol with the smaller particles has a higher viscosity for any given concentration.

A mathematical treatment of the viscosity of these systems was made by Einstein (*Ann. Physik.*, 1906, **19**, 289), who arrived at the expression

$$\eta_s = \eta_0 (1 + 2 \cdot 5\phi)$$

where η_s is the viscosity of the disperse system, η_0 that of the dispersion medium, and ϕ is the ratio of the volume of disperse phase to the volume of the sol.

According to this formula the viscosity should vary linearly with the concentration of disperse phase and should be independent of the size of the particles since no dimensional term referring to the particles appears in the equation. The validity of the equation has been established for particles of microscopic size and at low concentrations, but for sols and especially at higher concentrations it appears from the work of Odén just quoted that the increase of viscosity is more rapid than linear and that the viscosity also increases with decreasing size of particle.

One probable reason for this discrepancy is that the value of ϕ in the equation is not what we take it to be. The ratio of the *volume* of disperse phase to the total volume of the system cannot be measured directly. What we measure is the *weight* of disperse phase and calculate its volume from its density in the massive state, assuming that the density does not change during subdivision to the minute size of colloid particles. Whether such a change does occur is not known in the first place and there is still another factor to be taken into consideration, for as pointed out at the beginning of this chapter, it is possible and in many cases very probable that the particles are associated with envelopes of the dispersion medium, which contribute to the volume. This effect is inconsiderable when the particles are comparatively coarse and it is in such systems that the equation is found to hold, but in highly disperse systems such an effect would make itself felt.

These features are particularly well illustrated by a recent investigation of titanium dioxide sols by Freundlich and Kross (*Kolloid-Z.*, 1930, 52, 37). The sol is positively charged and hydrophobic. The dispersion medium does not contain a detectable amount of titanium ions and none appears in dialysate, but it is probable that such ions are adsorbed on the particles because the sol as a whole gives a positive test with the usual reagents for titanium. Extraordinarily high concentrations (up to 64 per cent.) of titanium dioxide can be obtained in the sols. The viscosity increases with the concentration of the sol, the increase being more rapid at high concentrations; indeed, the relative viscosity of a 64 per cent. sol is 35. The surface tension of even the most concentrated sols does not differ from that of water. Density determinations indicate a value of 4·77 for the colloidal particles of titanium dioxide, whereas the density of the material in the massive state is only 3·8 to 4·3. The marked increase is considered to be due to the compression of the dispersion medium at the surface of the particles. The diameter of the particles is 20 to 60 $\mu\mu$ and they are probably not spherical.

CHAPTER XI
EMULSIONS

In this book colloids have been classified for convenience as hydrophobic and hydrophilic, the main criterion of which is stability, and also as suspensoid and emulsoid, the former having solid and the latter liquid particles for the disperse phase. Although the two systems are fairly coincident, there is very scanty evidence to show that the disperse phase of most emulsoid colloids is really in the form of liquid droplets and we have now to discuss true liquid-liquid emulsions which resemble the hydrophobic rather than the hydrophilic colloids in many respects. Some of these would undoubtedly be classified as hydrophobic in one scheme and emulsoid in the other.

True emulsions have been studied less than suspensoid colloids, but Clayton (*The Theory of Emulsions and Emulsification*, London, 1923) has brought together the available data. Milk is a typical emulsion, consisting of fat globules in water containing small quantities of other substances. The turbid condense water from steam engines is an emulsion of oil in water, the emulsions of most immediate interest contain water as one of the liquid phases, and these emulsions will be in general of two types—*oil-in-water* emulsions or *water-in-oil* emulsions—meaning as a rule by the term " oil " a liquid which is not miscible with water.

It is not true to say that which of these types of emulsion is formed from the two liquids depends entirely on the volume ratio in which the phases are present, although this is a factor of prime importance. The liquid present in greater amount does not always form the continuous phase. It is necessary to know how to distinguish one type from the other and this may be done by quite simple tests. An oil-in-water emulsion consists of minute droplets of oil dispersed in water as the continuous phase and a water-in-oil emulsion contains droplets of water as the disperse phase. One method of distinction is to add a few grains of a dye which is soluble in the oil, but insoluble in water ; an oil-in-water emulsion does not become coloured because the dye does not reach the oil droplets, but a water-in-oil system immediately becomes coloured. The electrical conductivity also gives a ready indication, because when water is the disperse phase the conductivity is considerable as a result of the presence of electrolytes in traces, whereas the conductivity is extremely small when the continuous phase is oil.

Stable emulsions containing about 0·1 per cent. of oil in water can be made by pouring a solution of the oil in alcohol or acetone into a large volume of water. The concentration obtained is comparable with that observed in the case of suspensoid or hydrophobic colloids and the researches of Donnan (*Z. physikal. Chem.*, 1899, **31**, 42), Ellis (*ibid.*, 1911, **78**, 321 ; 1915, **89**, 145), Powis (*ibid.*, 1915, **89**, 91, 179, 186) and others have shown that the general colloid properties of these oil-in-water emulsions are closely similar to those of the typical hydrophobic sols and that their stability is mainly influenced by their electric charge. The oil particles are in Brownian movement, they are negatively charged and are coagulated by electrolytes, the valency rule being again in evidence.

In some respects the dispersed oil particles resemble solid particles, in that they are retained by an ultrafilter ; this shows that the globules are not readily deformed so as to enter the capillaries of the membrane and that more rigidity must be ascribed to them than to the liquid in bulk. This increase in rigidity is proportional to the interfacial tension between the two phases, so that for the true liquid properties of the disperse phase to be apparent the interfacial tension must be low.

Returning to the effect of the ratio of the volumes of the two phases on the type of emulsion formed, Pickering (*J. Chem. Soc.*, 1907, **91**, 2002) found it possible to prepare oil-in-water emulsions (the water containing soap) in which the dispersion medium occupied only 1 per cent. of the total volume. Obviously, the phase ratio cannot alone determine the type of emulsion produced. The matter becomes clearer by the following considerations. Suppose the disperse particles to be spheres of equal diameter and non-deformable, then we know from mathematical theory that the closest packing is when each sphere is touching twelve other spheres and that at this stage the spheres occupy about 74 per cent. of the total volume. The phase ratio is then 74/26, and this is the highest concentration possible, provided the system has the characteristics mentioned. When, however, the disperse phase is liquid the globules are merely touching when the phase-ratio is 74/26 and further increase in the phase-ratio can be effected by deformation of the particles by flattening at the twelve points of contact so that ultimately a regular dodecahedron would be formed. There is therefore theoretically no upper limit to the phase-ratio in pure liquid-liquid systems, but it must be remembered that the development of polyhedral shape from the spherical shape involves an increase in surface which is opposed by the interfacial tension, so that in practice a high percentage of disperse phase is only possible when the interfacial tension between the two phases is low.

This brings up another point of great importance. The discussion has referred to emulsions of two pure liquids, but as a rule it is not possible to get emulsions containing more than a fraction of 1 per cent. of disperse phase except in the presence of a third substance which is known as an *emulsifier*. The alkali soaps are particularly useful for this purpose. This factor introduced by the emulsifier not only determines what concentration of emulsion is possible, but also which type of emulsion is formed and is more important in the latter respect than the phase-ratio. In general, the alkali soaps produce oil-in-water emulsions and the soaps of the alkaline earths, iron, aluminium, and zinc favour the formation of water-in-oil emulsions. In fact, a reversal of type can be effected by adding barium chloride to a system stabilized by a sodium soap. For example, olive oil is readily emulsified when shaken with a dilute solution of sodium hydroxide; some sodium oleate is formed by chemical reaction and this acts as an emulsifying agent with the production of an oil-in-water system. On adding barium chloride the phases are reversed and a water-in-oil system is formed. The water-in-oil systems are as a rule very viscous or plastic and of the consistency of butter.

In the systems containing a high percentage of the disperse phase, the dispersion medium must exist in the form of very thin films dividing the particles of the discontinuous phase. With a liquid having the high surface tension of water such a structure would be very unstable and for it to persist there must be present some substance which lowers the interfacial tension considerably. The soaps are well known to do this and it is obvious that one of the functions of the emulsifier is to lower the interfacial tension between the phases, and the associated phenomena of adsorption and peptization occur as well.

Although it is easy to see that a decrease in interfacial tension favours the increase of surface which is a necessary accompaniment of higher dispersion, the question why one phase should take on a convex and the other a concave surface and why these should be reversed in the case of different emulsifiers remains untouched unless further assumptions are made.

Finkel, Draper, and Hildebrand (*J. Amer. Chem. Soc.*, 1923, **45**, 2780) have put forward a very ingenious theory in attempting a solution of this problem. They consider the shape of the molecules of emulsifier and suppose that peptization of an oil drop by an alkali soap takes place in such a way that the hydrocarbon chain points towards the oil and the dissociating sodium atom towards the water. Since the sodium atom is broader than the hydrocarbon chain, the soap molecules must fit like wedges, and when a molecular layer of the closest packing is formed it must give rise to a globule of oil in water. On the other hand, when the emulsifier is a calcium soap it is supposed that the two hydrocarbon chains are attached to the calcium atom in the form of a wedge. If the hydrocarbon chains turn towards the oil phase and the zinc atoms towards the water phase, the closest packing will be produced when water globules form in oil and consequently a water-in-oil emulsion is formed. Support for the theory is found when the behaviour of the different alkali soaps is compared with the atomic volume of the respective alkali metals. The argument does not seem to be universally applicable, however.

As an objection to arguments based on the formation of unimolecular layers of the emulsifier it may be urged that finely divided solid substances, such as lamp-black, often are good emulsifiers. The type of emulsion formed depends on the angles of contact of the two liquids with the solid emulsifier; the liquid with the smaller contact angle becomes the disperse phase.

Stamm and Kraemer (*J. Physical Chem.*, 1926, **30**, 992) consider that insufficient attention has been directed to the mechanism of emulsification, which consists of (1) pulverization of both phases into lamellæ and drops and (2) coagulation or reunion of the drops of one phase. The function of an emulsifying agent is to protect the drops of one phase to a certain extent whilst permitting the coagulation of the other phase, which then forms the dispersion medium.

The wettability of the emulsifier by the two liquid phases is undoubtedly of importance. An emulsifier which is more readily wetted by oil than by water stabilizes a water-in-oil system, whilst an emulsifier which is more readily wetted by water produces an oil-in-water emulsion. Actually, the alkali soaps are wetted more readily by water and alkaline earth soaps are wetted more readily by organic liquids. Charcoal is wetted less by water than by organic liquids and therefore produces a water-in-oil emulsion.

The viscosity of emulsions increases rapidly with the phase-ratio.

CHAPTER XII
PROPERTIES OF HYDROPHILIC COLLOIDS

The hydrophilic colloids are more important as a class than the hydrophobic colloids, if only because they form the basis of organic matter. Among other important groups, they include the complex carbohydrates and proteins. Nevertheless, they are much less understood and this is partly due to the fact that the majority of them consist of complex organic compounds whose chemical constitution is still unknown, and also because many of the methods of investigation applicable to other colloid systems are no longer of service.

There are therefore very serious difficulties in working with hydrophilic colloids and one of the chief causes is that the physical properties of the disperse phase and the dispersion medium are not so different as in the hydrophobic sols. In the first place, the ultramicroscope, the most convincing of all instruments for colloid research, is almost useless when applied to the majority of hydrophilic colloids. This is not always due to a smaller particle-size, although no doubt in some cases the particles are amicronic, but rather to the small difference in the optical properties of the two phases. Obviously, if the optical properties of the disperse phase and the dispersion medium are the same there will be no scattering of light at the interface. In most hydrophilic colloids there is to be expected little Tyndall effect and a lack of definition in the ultramicroscope. Another respect in which the properties of the two phases generally approximate is in the density, and this renders centrifuging and sedimentation investigations very difficult. Hydrosols of the noble metals contain particles having many times the density of the dispersion medium, but the density of most of the hydrophilic colloids is not markedly different from that of water. This precludes another main line of investigation, although it may be noted that an ultracentrifugal method of colloid research has just recently been developed by Svedberg

PROPERTIES OF HYDROPHILIC COLLOIDS 155

to such an extent as to give important data concerning the molecular weight of the proteins.

Again, the fact that the hydrophilic sols are generally colourless makes direct observation of cataphoresis very difficult, and methods of analysis of many of these complex compounds have not been sufficiently developed to make them useful for the purpose. On the other hand, there is one physical property of hydrophilic systems which is more variable and characteristic than in the case of hydrophobic sols. This property is viscosity, and viscosity measurements form one of the strongest features of investigations in hydrophilic systems. The practical aspect of viscosimetry is discussed in Chapter VI.

Viscosity. Referring to Einstein's equation for the viscosity of a disperse system (p. 146), it is found in practice that

$$\eta_s = \eta_0 (1 + 2 \cdot 5\phi)$$

hydrophilic colloids in sometimes quite small amount give sols of high viscosity quite out of proportion to the value of ϕ in the equation, representing the ratio of the volume of disperse phase to that of the dispersion medium. For example, a sol containing 2/7 per cent. of agar is quite viscous, having a relative viscosity of 2·400 compared with water. According to the formula the viscosity should be

$$1 + 2 \cdot 5 \times 2/7 \times 0 \cdot 01 = 1 \cdot 007$$

There is thus a great discrepancy, and it obviously indicates that the volume of particles in the sol is greatly in excess of that substituted in the equation. (The specific volume of agar is taken as unity in the above equation.) The volume of the disperse phase must be much greater than the volume of dry agar, and the only simple way to explain this is to assume that the agar particles have taken up a large amount of water, so increasing their volume.

In this connexion it must be pointed out that when dry hydrophilic colloids are placed in water the general effect is that they swell to several times the initial volume by imbibition of the water. The swollen gel often disperses directly in water, sometimes at more elevated temperatures, and after that stage it is not easy to tell whether more water is taken up. The viscosity of the sol generally increases for some time after the dispersion, however, suggesting that the volume of the disperse phase is increasing, or that the particles are growing.

To account for the experimental viscosity of the agar sol mentioned in accordance with Einstein's equation, it is necessary for the total volume of the particles of the disperse phase to be 200 times that of the dry agar. This may at first sight seem improbable, but when it is remembered that a sol containing only 1/7 per cent. of agar sets to a loose jelly at the ordinary temperature it will be realized that agar can " bind " 700 times its own volume of water. The large increase in volume of the particles required does not then seem so strange. Whether the particles take up the water after the manner of a sponge or whether they become surrounded by a thick atmosphere of water molecules is not known with certainty, but the latter view appears incredible when account is taken of the large distance over which the particles would have to exert their action. It is true to say, however, that the high viscosity of the hydrophilic sols leads us to suppose that the particles are highly hydrated. This also renders it more evident why the properties of the two phases are not very different and gives ground for the supposition that the disperse phase of the sol is liquid, even though the dry substance (as in agar and gelatin) may be a solid.

According to the law of Poiseuille, the viscosity of a liquid is independent of the rate of shear or the rate of flow through a capillary. This relation holds for pure liquids such as water, but measurements of the viscosity of a large number of hydrophilic colloids by Ostwald (*Z. physikal. Chem.*, 1924, **111**, 62 ; *Kolloid-Z.*, 1925, **36**, 99) show that in these cases the observed viscosity decreases with increasing rate of shear. The same has been found by numerous other investigators, among whom may be mentioned Hatschek and also Freundlich. The Couette viscosimeter described on p. 81 enables the variation of viscosity with rate of shear to be studied by using different speeds of rotation of the outer cylinder. With the capillary viscosimeter a modified form of apparatus can be used in which the liquid can start from different heights, thus giving different rates of flow.

This peculiarity of the viscosity of many sols is not really understood at the present time, but it is customary to imagine the systems exhibiting it as having in addition to ordinary liquid viscosity an elasticity analogous to that of solids. Kruyt (*Colloids*, Chapman & Hall, 1927, p. 170) considers that this view is unfounded. He writes :

"If we investigate sols in properly constructed viscometers, carefully accounting for the changes which the sol undergoes as a result of irreversible and hysteresis phenomena, and, furthermore, if we take care to exclude sols that gelatinize, the results obtained are in complete accord with the law of Poiseuille. Moreover, there is a complete absence of elastic phenomena."

That is not the view which is generally held, however. Still, it is necessary to bear in mind that there are many ways of measuring viscosity and that these methods often give different results with colloidal solutions. The results may be affected in some cases by quite distinct processes, such as the adsorption of the colloid by the rotating cylinder in the Couette apparatus, thus increasing the roughness of the wall.

McBain (*J. Physical Chem.*, 1926, **30**, 239) believes that neither solvation nor the assumption of a film of solvent enveloping each particle is adequate as an explanation of the viscosity of colloids. The high viscosities are ascribed to the increased shear which is imposed on the solvent when the small, rigid colloidal particles cohere to form open or loose aggregates suspended in the solvent. These structures practically immobilize the solvent which they embrace, and this involves interference between the particles and deformation when the sol is made to flow. Neutral colloids are regarded as solvated micelles capable of aggregation, the linking up of the micelles to form larger structures being accomplished by forces as local and specific as those operating in adsorption or in residual affinity. The efficiency of a solvent will depend on the dismemberment brought about, and in the poorer solvents this dismemberment is less complete, with a resulting higher viscosity. This hypothesis of ramifying micellar aggregates has been adversely criticized by Hatschek (*ibid.*, 1927, **31**, 383).

The formula of Einstein would lead to the view that a simple linear relation is to be expected between relative viscosity and the concentration of a sol. It is generally found that the viscosity of hydrophilic sols increases more rapidly than the concentration. This might be explained by supposing that the hydration of the particles is a function of the concentration. The temperature coefficient of viscosity is much greater for hydrophilic sols than that of the pure dispersion medium, water.

The Electro-viscous Effect. A study of the effect of electrolytes on the viscosity of hydrophilic colloids of the type of gelatin and agar shows that the first small quantities of electrolyte added decrease the viscosity of the sol considerably, whilst further addition of electrolyte causes practically no more change. When the curves connecting measured viscosity with the concentration of electrolyte are plotted for a number of salts, some striking facts appear. The strongly marked influence of the first small amount of electrolyte recalls what we have formerly found in electro-capillary processes such as adsorption and coagulation. It is found also that the position of the curve is determined, in the case of agar for example, by the cation of the added electrolyte and that the anion has only a small influence. Since agar forms a negatively charged sol under ordinary conditions, it seems that adsorption of the ion of opposite sign is concerned in the process. Further, all the curves for ions of a given valency are nearly coincident, but the curves for ions of different valency differ considerably, the initial decrease in viscosity increasing with the valency of the ion. The small quantities of electrolyte involved cannot appreciably affect the hydration of the particles, and since the considerations mentioned above suggest an analogy with other capillary electrical processes, it seems reasonably probable that the viscosity of the sols is determined not only by the degree of hydration of the particles but also by the electric charge on the particles. It may be that in a charged particle the outer repelling sphere must be added to represent the effective volume of the particle, whilst when the particle loses its charge some or all of this volume is deducted. The increase of viscosity which is due to the electric charge of the particles is known as the *electro-viscous effect*.

Surface Tension. In marked distinction from the behaviour of hydrophobic colloids, the surface tension of hydrophilic sols is generally considerably lower than that of water. The lowering of surface tension produced by soaps of the alkali metals calls for special mention. Exactly what significance may be attached to this effect as a general phenomenon is not very clear, because in some cases at least it appears that the lowering is due to the presence of impurities in true solution. Gelatin sols, for instance, have usually a lower surface tension than that of water, but it is found that the best brands of gelatin, which contain few decomposition products, have little effect on the surface tension of water.

Foaming. Another characteristic of the hydrophilic type of colloids is their tendency to foam, even when the solution is extremely dilute in some cases. The colloid generally concentrates in the froth; it will do so if it lowers the surface tension of water. The connexion between surface tension and foaming power is not so simple as might be expected, however. It is often stated that foam is produced by those substances which lower the surface tension most. That is partly true, but substances which raise the surface tension of water also cause it to foam. A very conspicuous case is the dye Night-blue, which, although raising the surface tension of water, causes copious foaming.

Actually, it seems that the condition leading to the formation of foam is that the surface layer of the liquid shall have a different composition from the bulk of the liquid. This will happen if a substance in solution is either positively or negatively adsorbed at the liquid-air interface; that is to say, whether the surface tension is lowered or raised. The chief support for this view is that pure liquids do not foam, but that the presence of either molecularly or colloidally dissolved or dispersed matter confers the property of foaming, particularly in those cases where strong adsorption at the surface is known to occur. It offers a ready explanation of the fact that where strong positive adsorption occurs, as in the case of albumin solutions, a foam rich in the colloid is produced, the albumin being thus removed from the solution.

Some interesting views on film formation and foaming have been expressed by Foulk (*Ind. Eng. Chem.*, 1929, **21**, 815). Starting from the fact that in solutions the solute is either more or less concentrated in the surface layer than in the interior of the liquid, it is maintained that such a system will resist a force tending to bring about an equality of concentration, that is, it will resist a tendency to mix the surface layer with the rest of the liquid. The theory explains why the effect is the same whether the surface tension is raised or lowered by the dissolved matter. The view is expressed that film formation and film stability are separate questions, which have often been confused in the literature. The tendency of a liquid to foam is not easily measured and is often determined by noting the time taken for the foam to disappear. This procedure really affords a measure of the stability of the foam, which may have no direct connexion with the foaming power. In this connexion Foulk remarks, " Even man's social nature is influenced by foams ; to such an extent, indeed, that it sometimes demands a stable foam, as in beer and soda-water, and sometimes an instable one, as in champagne and ginger ale."

Foaming is also a matter of some industrial importance, because in many technical operations the presence of colloids causes troublesome frothing. In most cases, some means of diminishing or preventing the foam has been found and among the anti-foaming agents used, milk, butter, fats, etc., figure prominently. In most cases the substances used are effective only for that particular operation and are not universally applicable. For example, the foam on beer quickly subsides if a bottle of ether is merely unstoppered in its vicinity, but ether is powerless to reduce the foam of a glue solution. The investigation of the destruction of foam for industrial purposes has recently been undertaken by Meyer (*Farben-Z.*, 1930, 36, 481, 573, 619), who examined the effect of adding numerous organic liquids to a foam made with a glue solution. Fatty acids had more effect in destroying the foam than had the corresponding alcohols. No relation could be found between the surface tension of the liquid and its power for destroying foam. A mixture of linseed oil and alkali was found to be very effective in preventing foaming.

Stability. The impression has already been gained that the hydrophilic sols are the more stable class of colloids. With certain exceptions they are stable towards rise of temperature (albumin coagulates irreversibly), but even here difficulties immediately arise because a great number of these complex substances are slowly hydrolysed, especially on boiling, and on subsequent cooling it often must not be assumed that the sol contains the same substance or substances, but rather that it is mixed with their decomposition products. A gelatin sol, for example, after boiling contains decomposition products which lower the setting point of the gel formed on cooling. Towards lowering of temperature the typical behaviour is the setting of the whole liquid to a jelly and this is in most cases a reversible process, the sol being re-formed on subsequent warming, even though there may be a marked hysteresis in this process, as in the case of agar ; silicic acid is an example of the irreversible formation of a jelly.

HYDROPHILIC COLLOIDS

The effect of electrolytes is very characteristic, for small amounts which would coagulate hydrophobic sols do not coagulate the hydrophilic sols. It is true that concentrated solutions of electrolytes do flocculate hydrophilic colloids, but the quantity involved is so vastly different that the process is looked upon as something quite different and the term " salting-out " is used to distinguish this process from the effect of small quantities of electrolyte on a hydrophobic sol.

One of the big problems of colloid chemistry is to discover how far the same principles can be applied to the understanding of the two types of sols and in earlier times the stability of the two types appeared to be quite different matters. The particles of both types are readily shown to be electrically charged, but it is found that the charge on the particles of hydrophilic colloids varies not only in magnitude but also in sign with the hydrogen-ion concentration of the medium, being generally positive in acid solution and negative in alkaline solution. These facts, coupled with the consideration that proteins, which form the most important branch of hydrophilic colloids, contain in their chemical constitution both acid and basic radicals, have given rise to a treatment of this type of sol as systems of amphoteric electrolytes. The electrical double layer is therefore accounted for in the constitution of the particle itself.

Whilst there is much to be said for this point of view, a comprehensive treatment of hydrophilic colloids must include the complex carbohydrates such as agar and starch, where an electrolytic explanation on the same lines appears to be precluded. Agar and gelatin are very similar in colloid properties, though essentially different chemically. Unfortunately, to maintain the amphoteric electrolyte view of the behaviour of hydrophilic colloids destroys the idea of continuity between hydrophobic and hydrophilic sols. This might conceivably be avoided by endeavouring to treat the hydrophobic colloids also from the point of view of electrolytic dissociation, regarding the particle as a highly complex ion. In that case, coagulation and other colloid-chemical processes would come in the domain of the solubility product, but as far as we know these phenomena cannot be explained in terms of a solubility product in a way analogous to the classical theory for electrolyte solutions. The electrical capillary phenomena which have been described throughout this book provide at present the only reasonable explanation of the behaviour of hydrophobic colloids.

The fact that we can now begin to regard the stability of the two types of sols from the same point of view is due in no small measure to the work of Kruyt (*Kolloidchem. Beih.*, 1928, **28**, 1). One important difference, determining the angle from which stability must be viewed, is that the hydrophilic sols are not flocculated when the charge on the particles is removed ; they are as a rule less stable, but they are still sols. The electro-viscous effect shows an analogy with the behaviour of hydrophobic sols, but it is obvious that there must be some other stability factor operating in the case of hydrophilic colloids. The considerations outlined in Chapter X suggest that this second factor may be the hydration of the particles, and the stability of the hydrophilic sols may therefore be ascribed both to the electric charge on the particles and to the protective action of the sheaths of dispersion medium bound to each particle. This is largely assumption at present, but is strongly supported by experimental evidence that the sols are flocculated *when discharge and dehydration occur simultaneously.*

The study of the electro-viscous effect shows that a decrease in viscosity is observed on adding small amounts of electrolytes to hydrophilic sols, the stability of the sol is generally reduced, the particles lose part or all of their charge, but the sol still remains as such. Discharge alone does not destroy the stability. However, on the addition of alcohol to a discharged sol flocculation ensues. Alcohol is known to be a strongly dehydrating agent, presumably on account of its forming hydrates in aqueous solution. The maximum in the viscosity curve for alcohol-water mixtures of varying composition is explained in this way. Thus, coagulation of hydrophilic sols occurs when discharge is followed by dehydration.

It is interesting to inquire whether the same effect can be brought about in the reverse order, that is to say, what is the effect of first adding alcohol as a dehydrating agent and then a small quantity of electrolyte as a discharging agent. When alcohol is added to a sol of agar, for example, a small quantity has little effect, but when a concentration is reached where its dehydrating action becomes considerable the properties of the sol change to a marked extent. The viscosity decreases rapidly and the optical properties become those of a hydrophobic sol. In fact, dehydration of the particles effects the conversion of a hydrophilic into a hydrophobic sol. This is an observation of prime importance. In accordance with this, on adding a small quantity of electrolyte the sol, now having hydrophobic properties, is coagulated, because the other stability factor, the electric charge, is thereby removed. The argument that the two stability factors for hydrophilic colloids are the electric charge and the degree of hydration appears therefore on a sound basis.

Measurements have been made of the amount of alcohol required to precipitate a sol which has been partly or fully discharged, and the results are variable, depending on the charge retained by the particles. It appears, therefore, that just as coagulation of hydrophobic sols occurs when the charge is lowered to a certain critical potential, so in the case of hydrophilic colloids there is a critical potential and also a critical degree of hydration. These two factors are interdependent, the degree of hydration required for stability being less the higher the charge.

The dehydration of hydrophilic sols by alcohol may be followed by viscosity measurements of the sol in presence of increasing amounts of alcohol. A decrease in the relative viscosity indicates a lowering of the value of ϕ in Einstein's equation, or a decrease in the volume of the particles. By relative viscosity is here meant the viscosity of the sol plus alcohol compared with that of water plus alcohol of the same concentration, for the viscosity of water itself is changed considerably by the addition of alcohol. When such experiments are conducted and the decrease in relative viscosity of the sol is plotted against the percentage of alcohol, the curve shows a steep fall at about 45 per cent. of alcohol, showing that about this concentration alcohol begins to exert its strongly dehydrating action. Acetone is another suitable dehydrating agent and gives similar results.

Salting-out. As one of the characteristic differences between hydrophobic and hydrophilic colloids it has been mentioned that the latter, though not coagulated by small quantities of electrolytes, are thrown out of the dispersion medium when large quantities of soluble salts are added. This phenomenon is called " salting-out " and it finds a ready explanation when viewed in the light of the considerations concerning the two factors for stability just discussed. The addition of concentrated solutions of salts provides simultaneously both conditions for the destruction of the stability; it has both a discharging and a dehydrating effect, for there is no doubt that concentrated salt solutions " bind " water, possibly because of the formation of hydrates of the ions. On this view, salting-out by concentrated salt solutions is a process of dehydration accompanied by the conditions for discharge of the dehydrated particles. In practice, the most suitable electrolytes for salting out are the sulphates of ammonium, magnesium, and sodium, not only on account of their high solubility, but also for the specific reason that they contain the sulphate ion, a point which will be discussed in the next section.

The Hofmeister or Lyotropic Series. Salting-out does not depend merely on the solubility of the salt employed, but on the specific nature of the ions present. Taking a series of salts of the same metal, the anions can be arranged in a definite order of effectiveness for salting-out. This was discovered by Hofmeister (*Arch. exp. Path. Pharmakol.*, 1899, **24**, 247; **25**, 1) and the series of anions is therefore called the Hofmeister series. The following table is taken from some of the results of Hofmeister and gives the concentration in g. mols. per litre of a number of sodium salts required to salt out a sol of albumin.

HOFMEISTER SERIES.

Citrate	0·56	Nitrate 5 42
Tartrate	0 78	Chlorate . . . . . 5·52
Sulphate	0 80	Iodide ⎫ Saturated solutions
Acetate	1 69	Thiocyanate ⎭ do not salt out.
Chloride	3 62	

The anions near the beginning of the series are therefore the most effective for salting-out and that is why among the common acid radicles sulphates are very suitable for this purpose. The anions at the latter end of the series not only have little or no effect in salting-out, but generally exert the reverse effect; that is to say, they aid dispersion. Many normally insoluble substances, such as cellulose, are readily dispersed in concentrated solutions of iodides or thiocyanates. The observation has also been made that, so far as albumins are concerned, in an acid medium this series is reversed.

This series is not only encountered in the salting-out of hydrophilic colloids, but also in examining the effect of various salts on other physical properties of water, for example on the solubility of non-electrolytes, the rate of inversion of sucrose, the rate of hydrolysis of esters, the surface tension of water, and the displacement of the position of the maximum density of water (normally at 4°). The relations between these effects have been discussed by Freundlich, who has introduced the term *lyotropic series* to indicate the wider applicability. It should be mentioned that some investigators have expressed doubt as to the validity of such series and hold that hydrogen-ion concentration is in all cases the controlling factor and that when account is taken of the various hydrogen-ion concentrations of solutions of the different salts any such regularity disappears. This view is not generally held, however.

Gortner, Hoffman, and Sinclair (*Kolloid-Z.*, 1928, **44**, 97) have made a study of the peptizing influence of a number of inorganic salts on wheat flour, the salt solutions being used at concentrations of 0·5 N to 2·0 N. The results show that the effect of the anion increases according to the following series: fluoride, sulphate, chloride, tartrate, bromide, iodide. Other investigations have shown that the fluoride ion should be placed at the beginning of the lyotropic series, indicating high precipitating and low dispersing power. This is consistent with the generally anomalous behaviour of fluorides in relation to the other halides. The peptizing effect of the cation was found to increase in the order: sodium, potassium, lithium, barium, strontium, magnesium, calcium. Differences in hydrogen-ion concentration are not responsible for these results. With increasing concentration of the salt solution the peptizing power of the alkali halides decreases, but with the alkaline earth halides the peptizing effect increases with concentration.

It is well known that consideration of the mobilities of different ions leads to the view that these must be hydrated to different extents and a relation between the lyotropic series and hydration of the ions immediately suggests itself. If the series is an indication of the order in which the ions are hydrated, the effect of the salts on the solubility of other substances becomes intelligible, for the more water required by the ions the less is available as a solvent for other material. Similarly, the salting-out of colloids by dehydration and the alteration of the rate of hydrolysis of esters (by varying the active mass of the water) are readily understood, but ionic hydration alone is not sufficient to explain the change in the surface tension of water or the change in the position of its maximum density.

These two properties are connected with the molecular complexity of the liquid. The variation of the surface tension of water with temperature suggests that liquid water molecules are associated and contain at least such complexes as $(H_2O)_2$ and $(H_2O)_3$. X-ray spectrographic data for the ice crystal suggest a molecule of H_6O_3 at low temperatures and these molecules have the open structure characteristic of covalent linkages. It is probably the formation of these molecules that causes the peculiar expansion of water when lowered from 4°. The two properties under consideration are therefore closely related to the molecular complexity of water. The state of water can be represented by the equilibrium

$$nH_2O \rightleftharpoons (H_2O)_n$$

and the inference is that the equilibrium is altered by different ions to different extents. It is probable, therefore, that the lyotropic series expresses the order in which the ions take up water or affect the degree of association of water.

The Iso-electric Point. As the electric charge is not the main factor governing the stability of hydrophilic sols, it is capable of variation to an extent quite impossible in hydrophobic colloids. Cataphoretic experiments in acid and alkaline solutions show that the sign of the charge varies with the hydrogen-ion concentration of the medium. In general, the particles are positively charged in presence of acid and negatively charged in alkaline solutions, probably by adsorption of hydrogen or hydroxyl ions. Obviously there must be some intermediate hydrogen-ion concentration where the particles are not charged either positively or negatively. This hydrogen-ion concentration does not coincide with the point of neutrality of water and varies from colloid to colloid, being specific for each. The hydrogen-ion concentration at which the particles are uncharged is known as the *iso-electric point* of that colloid. This value is open to determination by cataphoresis, for at the isoelectric point the disperse phase does not move in an electric field.

Since at the isoelectric point one of the stability factors has been removed it will be expected that in this state hydrophilic

HYDROPHILIC COLLOIDS

colloids show a minimum of stability. This is true as a rule, but there are exceptions, such as silicic acid, which has a maximum stability at the isoelectric point. The following table shows the isoelectric points of some proteins and also the hydrogen-ion concentrations at which they are most easily precipitated, and the agreement is good.

ISOELECTRIC POINT OF PROTEINS.

Protein	By cataphoresis	By precipitation
Natural serum albumin	2×10^{-5}	—
Denatured serum albumin	ca. 4×10^{-6}	3.8×10^{-6}
Serum globulin	ca. 4×10^{-6}	3.8×10^{-6}
Casein	2×10^{-5}	2×10^{-5}
Oxyhæmoglobin	1.8×10^{-7}	—
Trypsin	3×10^{-4}	3×10^{-4}
Gelatin	2×10^{-5}	—

CHAPTER XIII
INDIVIDUAL HYDROPHILIC COLLOIDS

In the preceding chapter are discussed some matters which are referable to hydrophilic colloids in general. In other respects there is great variation in behaviour from one colloid to another and it is not possible to go a great deal further into the properties of these sols without treating each colloid individually. In the present chapter it is proposed to describe the properties of some of the most important hydrophilic colloids.

General Properties of Proteins. The proteins comprise the most complicated molecules we know, and whilst their study from the colloid chemical or any other point of view presents greater difficulties than are encountered in other groups of substances, it seems right to treat them first among the hydrophilic colloids because they are in reality the classical examples of colloids. Glue, which suggested the term colloid, is a mixture of proteins. These substances form the most important part of living matter and the strictly chemical aspect of their properties and behaviour is studied in works on physiological chemistry. It is not proposed here to treat the organic chemical aspects, but to discuss them only so far as colloid chemical principles are involved. They contain the elements carbon, hydrogen, oxygen, nitrogen, and sulphur and in individual cases other elements; their chemical constitution is so complex as to defy attempts to ascertain it at present and even the molecular formulæ are so complicated that elementary analysis often becomes a matter of difficulty. It is known that the amino-acids play an important part in their constitution, and the proteins are regarded fundamentally as complex polypeptides. By peptides are understood chains of amino-acids linked together by the group $- CO.NH -$. The simplest peptide is glycylglycine,

$$NH_2.CH_2.CO.NH.CH_2.COOH.$$

That groupings of this sort must occur in the complex molecules is clear from an examination of their degradation products. It is clear, also, that such bodies, which may for simplicity be represented by $R\begin{smallmatrix}\diagup NH_2\\ \diagdown COOH\end{smallmatrix}$ have both acidic and basic properties and will be able to form two types of ions according to whether they react with acids or alkalis; thus $R\begin{smallmatrix}\diagup NH_3{}'\\ \diagdown COOH\end{smallmatrix}$ and $R\begin{smallmatrix}\diagup NH_2\\ \diagdown COO^-\end{smallmatrix}$. There is also the possibility of neutralization of the basic and acidic portions taking place inside the molecule itself and producing a double ion which is on the whole externally neutral in charge; thus, $R\begin{smallmatrix}\diagup NH_3{}'\\ \diagdown COO^-\end{smallmatrix}$.

Considering the variety of behaviour open to such chemically complex substances, and considering also the difficulties of examination which have been mentioned, it can only be expected that views on the nature of the proteins are not yet stabilized. Matters will no doubt gradually become clearer with time, but at present it is necessary to bear in mind that most views on the subject are open to criticism by adherents of other opinions.

Broadly speaking, proteins may be considered in three groups. *Simple proteins* include albumins and globulins and the higher polypeptides in the narrower sense. They are without smell or taste, generally amorphous, though sometimes crystalline, and are weakly both basic and acidic. Besides such well known substances as serum-albumin and egg-albumin, the simple proteins include the globulins from plant and animal sources, fibrinogen and fibrin from the blood plasma of mammals, which by fibrin ferment are held to bring about the coagulation of blood, myosin and myogen, muscle protein, the coagulation of which under the influence of a ferment is probably the cause of *rigor mortis*, phosphorglobulins such as the casein of milk and the vitellin of egg-yolk, histones, basic proteins which are precipitated by alkalis, and protamines, which are free from sulphur but rich in nitrogen.

The *compound proteins* are compounds of proteins with other bodies such as nucleinic acid, which gives rise to the nucleo-proteins; these are constituents of the cell nuclei. To this class also belong hæmoglobins, which are compounds with hæmatin, and the glycoproteins, mucines, and mucoids, which are compounds of proteins with carbohydrates.

The albuminoids form the structural tissues of animals and differ from the simple proteins in their insolubility in water, salt solutions, and the animal fluids. They are relatively inert chemically. They include collagen, which is the fundamental organic constituent of bones and cartilage and which is degraded by prolonged boiling with water into glutin, gelatin, or glue; keratin, which is the chief constituent of hair, wool, feathers, skin, etc.; elastins, constituents of elastic tissues; fibroin, contained in silk; and spongin, which is found in the sponge.

The specific properties of the more important proteins will be discussed later, but some of the general effects observed may now be mentioned. On digestion in the stomach the proteins come into contact with an enzyme known as pepsin and with a solution containing about 0·4 per cent. of hydrochloric acid in the presence of which pepsin becomes active. Whether the proteins are in their natural state or are denatured as in the case of the white of a boiled egg, under the influence of the enzyme the structure is broken down in part forming primary and secondary albumoses and finally peptones. The albumoses diffuse very slowly and can be separated by ultrafiltration through a glacial acetic acid collodion filter into different fractions known as heteroalbumoses, protalbumoses, and deuteroalbumoses. Even these products are still mixtures and may be fractionated by alcohol. The albumoses are precipitated by ammonium sulphate solution. On the other hand, the peptones, the end products of pepsin digestion, are more similar to crystalloids. They diffuse easily, pass slowly through membranes, and are not precipitated by ammonium sulphate solution. Albumoses and peptones are not coagulated on boiling and differ in this respect from proteins. The peptones are mixtures which may be fractionated.

Leaving the stomach, the substances enter the intestines and encounter another enzyme known as trypsin, which is active in the presence of the alkali of the intestine; this effects further degradation into simpler amino-acids.

The proteins exert a measurable osmotic pressure and this is increased, in some cases investigated, by the addition of acids and alkalis. The work of Lillie (*Amer. J. Physiol.*, 1907, **20**, 127) may be mentioned in this connexion. It might be supposed that the increase in osmotic pressure of gelatin on adding acids or alkalis is due to hydrolysis to simpler molecules, but against this is the fact that the original value of osmotic pressure is restored when the acid or alkali is removed by dialysis. It seems, therefore, rather that the addition of these reagents brings about a subdivision of the gelatin particles. On the other hand, the addition of neutral salts serves to decrease the osmotic pressure of gelatin and albmumin sols, and here we must suppose that incipient coagulation or coarsening of the particles is brought about. It is interesting to note that when the anions of salts of the same metal are arranged in decreasing order of effectiveness in reducing the osmotic pressure of protein sols, the Hofmeister series is reproduced.

The general effect of salts of the alkali metals in precipitating proteins has been mentioned in the previous chapter, where the lyotropic series was introduced. Pauli has found that there is a relation between this precipitating effect on proteins and physiological action. Citrates, tartrates, and sulphates, at one end of the series excite the intestines and raise the blood pressure, whilst nitrates, bromides, and thiocyanates at the other end lower the blood pressure. As to the action of salts of the heavy metals on protein solutions, as a rule smaller amounts are required for precipitation and the process is often irreversible in the sense that on washing out the electrolyte the sol is not reproduced. Sometimes, however, the precipitate is soluble in excess of the precipitating agent.

Some of the individual members may now be considered.

Gelatin. Gelatin is a mixture of proteins, the principal constituent being glutin, and is made by boiling bones, hoofs, etc., in water to which acid has been added. Its elementary composition is somewhat variable, as shown by the following approximate analyses.

Element.	Percentage.
C	49–50
H	6 4–6·7
N	17·8–17·9
S	0·2–0·7
O	24–25

The most variable constituent is sulphur, suggesting that some substance rich in sulphur is present as an impurity. There is also generally a mineral ash content, and in some brands this is considerable.

Commercially, gelatin is sold in the form of air-dry horny sheets and also in a powder form. Since no two brands are exactly similar it is usual in experiments not only to state the source of the material, but also to take a sufficiently large stock to last through all the experiments. To prepare a sol, the sheets are broken into small pieces, put into a beaker, the requisite amount of water poured on, and the gelatin left for some hours, during which time it swells considerably, taking up about ten times its own weight of water. The contents of the beaker are then warmed at 35–40° in a water bath and the gelatin disperses to a sol. If it is to be kept for any length of time some agent such as thymol must be added to prevent bacterial decomposition. The various properties of gelatin described later are affected very much by all the details of the method of preparation of the sol and it is therefore essential to carry out exactly the same procedure as far as possible in every case. Small factors producing marked effects on some of the properties of the sol, apart from the presence of impurities, are the time allowed for swelling of the gelatin, the temperature of dispersion, the temperature to which the sol has subsequently been heated, and the rate of cooling. For many purposes it is desirable to remove the electrolyte impurities which are present in commercial gelatin. The ash is generally rich in calcium salts, and the presence of chloride, sulphate, phosphate, and carbonate can generally be detected. These can be very largely removed by soaking the gelatin leaf in distilled water which is either kept running or is renewed at frequent intervals for a period of two days or so. It should be pointed out that the properties of the gelatin are appreciably affected by this treatment, but it is advisable in some work. As the operation takes some time, it is best to place a small piece of thymol in the beaker to prevent putrefaction.

The osmotic pressure of gelatin has been measured and is found to increase with rise of temperature at a greater rate than is characteristic for crystalloids. On cooling, the original osmotic pressure is not restored until after some time. These observations suggest a greater dispersity of the gelatin particles at the higher temperature. In the ultramicroscope, the warm solutions appear to be homogeneous and whether or not particles can be observed in the cooled sol depends on the concentration. When this is between 0·1 per cent. and 6 per cent. particles may be observed, but at higher or lower concentrations only amicrons are present.

Two of the most interesting properties of gelatin are the swelling which occurs on placing in water or salt solutions and the setting of the sol to a jelly on cooling. These phenomena will be discussed in some detail in Chapter XV. The temperature of gelatinization depends on the concentration of the sol, and at room temperature a concentration of about 1 per cent. is necessary. This does not give a very firm gel, but 5 per cent. or 10 per cent. gelatin gels can easily be handled. The gelatin gels are very elastic.

When gelatin sols are boiled they no longer set to a gel when cooled to room temperature and the non-gelatinizing product is known as β-gelatin or gelatose. In the ordinary way small amounts of this transformation product are present in gelatin or play a large part in determining the properties of different samples. Gelatins are classified as " hard " and " soft." The former contain little gelatose, the gels have a relatively high melting-point and are firmer, whilst the soft gelatins contain a higher percentage of gelatose and a higher concentration is necessary to form a gel at a given temperature. On prolonged boiling, and more readily in the presence of acids, albumoses and peptones are formed from gelatin.

Ordinary salt solutions do not precipitate gelatin unless high concentrations are employed, when salting-out occurs. Tungstic acid is a good precipitant for gelatin; it is also precipitated by tannic acid in the presence of salts and by mercuric chloride in the presence of sodium chloride and hydrochloric acid.

Gelatin is remarkable for the highly protective effect it exerts on hydrophobic colloids.

The molecular weight of gelatin has long been a matter of interest. Eggert and Reitstötter (*Z. physikal. Chem.*, 1926, **123**, 363) deduced from measurements of the osmotic pressure of electrodialysed gelatin the value 40,000, commercial gelatins giving values round about 30,000. Wintgen (*Kolloid-Z.*, 1929, **47**, 104) has studied the precipitation of gelatin by sols of ferric, chromium, and aluminium hydroxides. The experiments with ferric and chromium hydroxides agree in yielding the value

30,000 for the "equivalent aggregate weight" of gelatin, thus in good agreement with the values determined from osmotic pressure measurements. The experiments on precipitation with colloidal aluminium hydroxide gave a value of only 19,700, but since the intermicellar liquid was found to contain basic aluminium chloride, it is believed that the latter has a dispersing effect on some of the gelatin, thus reducing its apparent molecular weight. The equivalent aggregate weight of gelatin is greater at the isoelectric point and alters with time, undergoing a fairly sharp rise to about 220,000, followed by a slow fall to about the original value. Increasing quantities of alkali progressively diminish the equivalent aggregate weight of gelatin.

The sedimentation investigations of Svedberg (described on p. 177) have shown that gelatin sols are polydisperse, containing particles leaving molecular weights between 10,000 and 70,000.

Albumins. The albumins are perhaps the most important of the hydrophilic colloids because of the part they play in the animal organism. Much research has been done on them and a good deal is already known about them, but a vast amount remains to be discovered and this will occupy biochemists for some time to come. The description of these substances properly belongs to works on physiological chemistry and here it is proposed to give only the barest outline.

Albumins are richer in sulphur than other proteins, containing from 1·6 to 2·2 per cent. The best known, egg albumin, occurs in the white of egg and can be obtained therefrom by precipitating the globulin and ovomucoid also present by adding a half-saturated solution of ammonium sulphate. After removal of the precipitate, the ammonium sulphate is removed by dialysis of the clear liquid and a sol of albumin is left. Generally the starting point will be dried egg albumin which is sold in the form of a powder. To prepare a sol, the powder is poured slowly into the requisite amount of water with constant stirring. The turbid sol should then be filtered through glass wool and the filtrate should be merely opalescent. The sols easily putrefy and a little thymol may be added to prevent this.

A characteristic feature of the albumin sol is that, although it does not gelatinize on cooling, it undergoes an irreversible coagulation when heated to about 60°, with formation of the white substance familiar in cooked eggs. This change is favoured by the presence of salts. It seems that this process consists of two stages, the first a denaturation of the protein and the second a coagulation of the denatured protein. The first process probably involves a chemical change. The dual nature of the process is shown by some interesting work of Pauli and Handowsky, who found that iodides and thiocyanates prevent the coagulation by heat. However, on dialysing away the iodide or thiocyanate coagulation occurred immediately, showing that denaturation had occurred and that the salt only prevents the subsequent coagulation of the denatured albumin.

The heat coagulation of serum-albumin has been studied by Wilheim (*Kolloid-Z.*, 1929, **48**, 217). Serum-albumin which has been coagulated by heat can be brought into solution by treatment with concentrated solutions of thiocyanates, salicylates, and benzoates. After dialysing away the salt, the albumin is again coagulable. The salts concerned hinder the coagulation by heat. Two essential conditions for this dissolution are a minimal concentration of the salt and a sufficient amount of liquid, the latter depending on the amount of albumin coagulum. The dissolution is therefore supposed to occur in two stages: in the first stage the coagulated albumin swells, this being controlled by the concentration of salt, and in the second stage the swollen albumin is dispersed, according to the amount of liquid available. Many other salts produce this effect when in sufficiently high concentration and the anions increase in effectiveness according to the lyotropic series. Since the process of dissolution appears to be definitely a swelling phenomenon, it would seem that the coagulation by heat is due to an energetic loss of water, and this view is supported by the fact that lithium halides, which have a strong swelling effect, also dissolve the albumin coagula. The solutions of albumin in alkali thiocyanates and salicylates are salted out with difficulty, even when the electrolyte has been dialysed away; such solutions are precipitated by acids. Measurements of the viscosity of the solutions with different concentrations of thiocyanate or salicylate in the cold show only a slight and variable increase in viscosity, but if the solution be boiled the viscosity-concentration curve rises very steeply, passes through a maximum, and then falls to the value for the unheated solution. This behaviour is probably connected with alterations in the degree of dispersion. The osmotic pressure is reduced after boiling the solution.

Salts of the alkali metals and of magnesium salt-out egg albumin only in high concentrations and the coagulum is reversible; salts of the alkaline earths salt out in high concentrations, but the coagulum becomes irreversible when left to stand; salts of the heavy metals salt out irreversibly at lower concentrations, but in many cases the coagulum redissolves or is peptized by more concentrated solutions of the salt.

Globulins. The globulins are not soluble in pure water, but dissolve in dilute neutral salt solutions. They are precipitated by concentrated solutions of sodium chloride or magnesium sulphate and by half-saturated solutions of ammonium sulphate. This last method is used in the separation of egg-globulin from the white of eggs. Ammonium sulphate is added until the solution is half-saturated and the globulin is precipitated. The filtrate contains albumin and ovomucoid. Globulin is also precipitated when the white of egg is dialysed against distilled water. Normally, there is sufficient electrolyte in the egg-white to keep the globulin in solution.

Hæmoglobin. Hæmoglobin is the colouring matter of the red corpuscles of blood. The liquid in these corpuscles is isotonic with a solution of sodium chloride containing nine parts in a million of water, so that if the cells are put in a more dilute solution they burst and the hæmoglobin goes into the outer solution. The solution is definitely colloidal in its properties, but crystals may be obtained from it. The crystals of hæmoglobin from the blood of different animals do not have quite the same composition, but the molecule is peculiar in containing iron in its constitution. The formula $C_{758}H_{1203}N_{195}O_{218}FeS_3$ has been given for the hæmoglobin of the blood of the dog.

The most striking property of hæmoglobin is to take up oxygen and give it up again easily and this is its function in the blood circulation. The product of oxygen absorption is known as oxyhæmoglobin; oxygen is taken up by the blood in the lungs and transported in the form of oxyhæmoglobin to the tissues where oxidation processes are required.

Carbon monoxide is also readily absorbed by hæmoglobin and the resulting bright-red substance is called carbon monoxide-hæmoglobin. From this compound the carbon monoxide is given up with some difficulty and carbon monoxide can readily displace oxygen from oxyhæmoglobin. This accounts for the poisonous effect of carbon monoxide.

The molecular weight of hæmoglobin has been approached by several chemical methods. It is found that 1 grm. of hæmoglobin combines with 0·00167 grm. of carbon monoxide under normal conditions and assuming that simple molecular association takes place the molecular weight of hæmoglobin is calculated to be 16,720. Alternatively, assuming the hæmoglobin molecule to contain one atom of iron, the molecular weight works out to be 16,660, which is in excellent agreement with the other value. There is, of course, an assumption in each of these methods which has nothing to support it, and if the hæmoglobin molecule contains n atoms of iron it must combine with n molecules of carbon monoxide. Osmotic pressure determinations have given values agreeing satisfactorily with those quoted above.

Casein. Casein is a phosphorus-containing protein and is the most important protein in milk (see p. 258), in which it exists in combination with calcium phosphate and from which it may be precipitated by the addition of acid. It is insoluble in water and in solutions of most neutral salts, but is soluble in alkalis and is sufficiently acidic to expel carbon dioxide from carbonates. The precipitate produced by acids redissolves in excess of the acid, an indication of the amphoteric nature of this substance. Casein solutions containing calcium are coagulated by an enzyme from the stomach of the calf known as rennet and this process is of importance in the making of cheese. In the absence of calcium, rennet does not coagulate casein sols, but some change must occur because subsequent coagulation occurs on adding a calcium salt.

Molecular Weight of Proteins. The determination of the molecular weight of proteins has long been a problem of outstanding difficulty. Proteins have resisted colloid chemical investigations, not only because of their unfavourable optical properties, but also because sedimentation methods of investigation have been difficult on account of the high dispersity and low density of the material. Nevertheless, some recent pioneer work by Svedberg (*Kolloid-Z.*, 1930, 51, 10) has aimed at studying the problem from the point of view of sedimentation and has given results already of the highest importance.

Refined centrifugal methods, using high speeds of revolution, have been applied to this problem, especially to the determination of the distribution of particle size. With hæmoglobin the sedimentation in the ultracentrifuge is very small and particle-size determinations gave the unexpected result that the system is isodisperse, the particles having a molecular weight of 68,000 and thus containing four iron atoms to the molecule (cf. p. 177). Experiments with serum globulin point to a molecular weight of 103,800 and for purified egg-albumin 34,400. The hæmocyanin of the snail is found to give an isodisperse system, although the molecular weight is about 5,000,000. The molecular weight indicated for serum-albumin is 67,500, which does not agree with Sörensen's value of 45,000, derived from measurements of osmotic pressure, but Svedberg rightly points out that in osmotic methods of determination the materials often suffer chemical degradation. The molecular weight may also change considerably with the hydrogen-ion concentration of the medium. For example, the hæmocyanin of the snail is stable between p_H 7·3 and 4·3, but outside this region the molecular weight falls from 5,000,000 to 100,000.

Gelatin is an example of a polydisperse protein system, the particle weight varying between 10,000 and 70,000. The diversity in the properties of different gelatin solutions of even the same concentration must be ascribed largely to this cause, for the degree of dispersion and the distribution of particle size will be affected by such factors as the time allowed for swelling and the temperature at which the swollen gel is dispersed to a sol. The monodisperse proteins include egg-albumin, hæmoglobin, serum-albumin, serum globulin, amandin, edestin, excelsin, legumin, C-phycocyan, R-phycocyan, R-phycoerythrin, H-hæmocyanin, L-hæmocyanin, and the polydisperse proteins include euglobulin, fibrinogen, gelatin, gliadin, globin, glutenin, histone, casein, lactalbumin, legumelin, leucosin, muscle-globulin, ovoglobulin, and pseudoglobulin. Members of the latter class are unstable and their dispersity changes with time.

An exceedingly important result of these investigations is the discovery that the monodisperse proteins can be arranged into two classes, having molecular weights of 35,000–210,000 and of several millions, respectively. The hæmocyanins form the latter class. Further, it is found that the first class can be divided into four groups, the molecular weights of which are respectively approximately 1, 2, 3, and 6 × 34,500. This regularity cannot be explained at present, but a simple aggregation hypothesis is ruled out. When the more complex proteins are degraded chemically they break up into particles of molecular weight 34,500, or a small multiple of that value.

The relation of the physical to the chemical molecular weight is not clear, nor is the state of polymerization of the chemical molecule apparent in these complex organic compounds. In some cases the empirical formula is quite simple, and Meyer (*Kolloid-Z.*, 1930, **53**, 8) has expressed the view that the polymerization is due to the action of intermolecular cohesive forces. On the other hand, Staudinger (*ibid.*, 1930, **53**, 19) supposes that in most of these colloids the particle is a single molecule, all the atoms being joined by ordinary covalent linkings, and not an associated group of units joined together by a separate cohesive force. He recognizes three kinds of organic colloids: (1) suspensoids and emulsoids, where the disperse phase is completely insoluble in the dispersion medium; (2) association or micellar colloids, which are heteropolar compounds giving an ion of colloidal dimensions; and (3) molecular colloids, which are constituted in the way described above.

Hydrosols of Carbohydrates

The simple sugars dissolve in water to form molecularly disperse solutions which exert the osmotic pressure and have all the other characteristics normally associated with true solutions. The size of these molecules is not very large and the molecular weight can be determined by the ordinary methods applicable to dilute solutions. There are also complex carbohydrates, such as cellulose, starch, dextrin, etc., which disperse in water under the influence of peptizing agents or in some cases spontaneously to give typically colloidal solutions, having a small, though sometimes measurable, osmotic pressure. The problem attached to these substances is very similar to that of the proteins; their chemical constitutions and molecular weights are unknown, being generally represented by the formula $(C_6H_{10}O_5)_n$, where n is unknown, and it is not at present clear whether a distinction should be made between the chemical molecule as the unit of chemical reaction and the physical molecule as the unit of separate physical existence. Some are of opinion that the chemical molecule is large enough to be of colloidal dimensions and that the solutions are truly molecular, although colloidal, whilst others believe that the particles in the sol are aggregates of molecules just as the particles of gold sols are aggregates of gold molecules. In any case, the question is complicated by polydispersity, but we may hope to gain some light on this aspect by the application of more refined sedimentation methods, such as have recently been applied to the proteins. These investigations do not lead, however, to the chemical molecule in its sense of unit reactivity, so co-operation in other directions is essential. Some assistance has already been derived from X-ray spectrographic data. X-ray examination of many natural fibres, such as those of cotton, indicates that these contain oriented particles arranged in a crystalline space lattice. From the diffraction patterns it is possible to gain an idea of the size of the molecule, and the work which has been done so far tends to show that this is very much smaller than the size of the dispersed particles in the sol. Probably the value of n in the formula given to the complex carbohydrates is much less than has been supposed, and there is a growing opinion that the colloidal particle of a hydrophilic sol may contain some thousands of molecules.

A few examples of the more important carbohydrate sols may now be described.

Starch. Starch occurs in the form of grains in many plants. It cannot be regarded as a definite colloid chemical individual, for the quantitative properties vary with the source of the material. Starch contains a complex carbohydrate known as amylose, and also amylopectin, to which is assigned the formula

$$R.CH_2O-P\begin{smallmatrix}\nearrow O \\ \leftarrow OH, \\ \searrow OH\end{smallmatrix}$$

where R is a carbohydrate radical. The starch grains are not peptized by cold water, but are at 60°, when a paste is obtained. When the paste is heated with water in an autoclave at 120° a very viscous sol is obtained. The viscosity of the sol diminishes with age, ultimately reaching that of water, and is also lowered by acids and is raised by alkalis. The initial high viscosity is probably due to the amylopectin, which is hydrolysed slowly, corresponding with the decrease in viscosity.

Starch grains can be peptized in the cold by alkali solutions and by solutions of many very soluble salts, such as the thiocyanates, potassium iodide, zinc chloride, etc., and the resulting sols are insensitive to electrolytes. The particles are too small to be resolvable in the ultramicroscope, especially considering that their optical properties are not sufficiently divergent from those of the dispersion medium, and they exert a small osmotic pressure, measurements of which indicate a molecular weight of about 100,000. A degradation product, known as soluble starch and having an apparent molecular weight of about 50,000, is formed when starch is warmed with hydrochloric acid.

Ostwald and Frenkel (*Kolloid-Z.*, 1927, **43**, 296) have made a study of the formation of starch paste from starch suspensions under the influence of various added substances. The substances which facilitate the change are sodium, potassium and ammonium thiocyanates, sodium and potassium hydroxides, hydrochloric acid, and carbamide. The viscosity curve of the process is S-shaped, like that of the setting of Plaster of Paris. The addition agent becomes effective at a certain critical concentration and has a large " concentration coefficient." The concentration of the starch suspension and the temperature are also of great influence on the velocity of formation of paste. Different velocities were found for different kinds of starch, and thus this effect can serve as a means for the characterization of starch.

The blue coloration formed by starch with iodine in presence of iodides is a well-known reaction and is used in volumetric analysis for determining the end-point of iodometric reactions. The nature of the substance formed has long been a matter of conjecture and dispute. An investigation by Angelescu and Mircescu (*J. Chim. Phys.*, 1928, **25**, 327) indicates that most of the iodine is taken up by the starch in the form of potassium tri-iodide. The blue colour formed is due to free iodine either directly adsorbed or liberated from the adsorbed potassium tri-iodide, and is not due to chemical combination between iodine and starch. Under certain conditions a red or violet colour is produced, and observations on the influence of electrolytes on the colour changes lead to the conclusion that substances which increase the degree of dispersion of the starch tend to produce a red colour. The colour changes are interpreted in a similar manner to the colours of colloidal gold solutions.

Lottermoser and Ott (*Kolloid-Z.*, 1930, **52**, 138) have directed experiments at a distinction between chemical reaction and adsorption processes in the formation of the so-called starch iodide. Measurements were made of the distribution of iodine between carbon tetrachloride and an aqueous potassium iodide solution containing starch, and potential measurements were carried out to study the adsorption of the different molecular species in the polyiodide system. The I_3' ion was most strongly adsorbed by the starch. The conclusion reached from these and other experiments is that the taking up of iodine by starch is in the first place an adsorption process and can be expressed by the ordinary adsorption isotherm.

Cellulose. Cellulose is a complex carbohydrate of the greatest importance, since it is the chief constituent of cotton, paper, and other woody matters. Chemically, it is generally represented by the formula $(C_6H_{10}O_5)_n$, but its true chemical composition and its molecular weight are not known. X-ray examination indicates that the unit cell contains four $C_6H_{10}O_5$ groups, but this gives no information about the size of the colloid particle.

Microscopic and ultramicroscopic examination of cellulose has shown the fibre to consist of a number of concentric tubes. The outer hard cuticle or skin consists of a mixture of waxy substances the chemical nature of which has not been clearly defined. Then follows a transition layer of cuticulated cellulose, a tube chiefly of cellulose, a second transition tube of plasmated cellulose, and finally an inner tube, which contains the central canal or lumen and represents a plasmatic formation. The cellulose tube consists of a number of very thin concentric tubes the walls of which are of unequal thickness and consist of ultra-microscopically thin fibres arranged parallel to each other along the length of the tubes.

Although cellulose does not swell markedly in pure water, it does so in electrolyte solutions and during the process the thickness of the fibres increases considerably, whilst the length of the fibre actually undergoes a decrease. This behaviour is consistent with the spiral formation of the structural units of the thread.

COLLOIDS

So far as the evidence goes, the swelling of cellulose appears to be connected with the formation of compounds between the cellulose and the electrolyte. Some interesting work on the swelling of cellulose in sodium hydroxide solutions was conducted by Pavlov (*Kolloid-Z.*, 1928, **44**, 44), who also measured the amount of sodium hydroxide taken up by the material. The swelling isotherms were found to exhibit a maximum and a minimum. At the swelling maximum, the true adsorption isotherm shows the taking up of one molecule of sodium hydroxide for each molecule of cellulose, and it is concluded from the results that the compounds $C_6H_9O_5Na$, $C_6H_8O_5Na_2$, $C_6H_7O_5Na_3$, and $C_6H_6O_5Na_4$ exist in a state of electrolytic dissociation in sodium hydroxide solutions of various concentrations.

Cellulose is not peptized in water alone, but is readily dispersed when boiled in concentrated solutions of very soluble salts. This dispersion or dispergation has been studied in detail by von Weimarn (*Rep. Imp. Ind. Res. Inst.*, Osaka, 1925, **6**, 37, 47, 63, 71), who found that rise of temperature favours the process and that the activity of the electrolytes used decreases in the order: calcium, strontium, barium; thiocyanates, iodides, bromides, chlorides. With less favourable salt solutions more drastic conditions have to be employed. Cellulose commences to be dispersed by solutions of sodium chloride or barium chloride when boiled with the saturated solution at 170° and 8 atmospheres pressure. At higher temperatures, some hydrolysis occurs simultaneously, forming a yellow or brown colloidal solution. By similar treatment with saturated lithium chloride solution at 160°, a maximum concentration of 10–12 grm. of cellulose per 100 c.c. can be obtained. Pure cellulose has two stability maxima —one in the presence of small amounts of electrolytes and the other in concentrated salt solutions. Intermediate concentrations of electrolytes precipitate the cellulose.

The dispersion of the natural fibre in concentrated salt solutions is impeded by the waxy coating of the material. During dispersion the fibre is divided into long fibrils, which ultimately break up along the length of the fibre.

Cellulose disperses readily in Schweizer's reagent (ammoniacal cupric oxide). According to Neale (*J. Text. Inst.*, 1925, **16**, 363 T) the cuprammonium cellulose solutions belong to the class of colloidal electrolytes. The strong base, cuprammonium hydroxide, forms with the weak acid, cellulose, a soluble basic salt having a crystalloidal cation and a colloidal anion. Each cellulose hexose unit is associated with one atom of copper and cellulose neutralizes the cuprammonium hydroxide to the extent of using completely the hydroxyl arising from the first stage of the dissociation of the base. A recent investigation by Stamm (*J. Amer. Chem. Soc.*, 1930, **52**, 3047) shows the cellulose-cuprammonium complex to be monodisperse with a molecular weight of 55,000 ± 7000, which on a copper-free basis becomes 40,000 ± 5000.

The cellulose can be precipitated from these solutions by the addition of acid. The micelles of both these and other dispersions of cellulose probably contain degradation products, for the regenerated substance has different properties.

These sols and the regenerated products are very important industrially, for they form the basis of the manufacture of artificial silk and other products. There are several processes. The viscose solutions are important. Mercerized cellulose (cotton treated with alkali) on treatment with carbon disulphide liquefies to a yellow-brown slimy mass, which contains cellulose xanthate and is readily dispersed by shaking with water. The sol is strongly alkaline and on neutralization with acid the cellulose is regenerated and may be pressed into a solid block. Many small fancy articles are made of this material.

In the manufacture of artificial silk fibres the viscose is squirted through a fine orifice into a coagulating solution such as dilute sulphuric acid. The sol then coagulates to a fine thread. Both the concentration and the age of the sol are important in this process. Probably, with age the micelles become rod-shaped and the squirting produces an orientation which enables the particles to cohere better.

Cellulose acetate and nitrate are important commercial products. Further reference to them is made on p. 192.

Agar. Agar is a substance of gelatinous appearance and properties, which is extracted from Chinese seaweed. It is mainly a mixture of complex carbohydrates of which the chief constituent is *d*-galactan. On hydrolysis with dilute acids the main product is *d*-galactose, with other hexoses and some pentoses. There is also present a protein-like substance containing nitrogen, which, however, appears to have little effect on the properties of the substance. In addition, there is a high ash content, a considerable proportion of which cannot be removed by dialysis. The ash consists chiefly of calcium sulphate, with a small amount of magnesium sulphate and traces of other salts and silica. An investigation by Fairbrother and Mastin (*J. Chem. Soc.*, 1923, **123**, 1412) leads to the conclusion that the chief constituent of agar is a sulphuric ester of calcium of the type $(R.O.SO_2.O)_2Ca$. By immersing the substance alternately in water and in dilute hydrochloric acid the free acid can be obtained; this is a gel, which hydrolyses very readily so that once it has been melted it will not set again. The potassium salt of agar forms a very stiff gel and can be made by treating agar with potassium oxalate. The dialysed product gives an ash of potassium sulphate which is approximately equivalent to the amount of calcium sulphate obtained from the original agar.

Agar is a familiar substance in the colloid laboratory and is usually sold in the form of shreds or in powder form. To make a sol, the air-dry agar is put into the requisite volume of water and allowed to swell therein for twenty-four hours. After swelling, the agar disperses at boiling temperature. The contents of the vessel must be stirred continuously and efficiently during the boiling in order to avoid cracking of the glass. The sol is then filtered, while still hot, as rapidly as possible through glass wool. Agar sols and gels are not liable to putrefaction and this fact coupled with the higher melting-point of the gel make it very useful for a number of investigations.

When the sol is cooled to about 35° it sets to an elastic and very firm gel, but there is a very marked hysteresis in this sol-gel transformation, for the gel does not redisperse to a sol until the boiling-point is almost reached. The actual setting-point depends on the concentration of agar, and this substance is noted for the rigidity of the gel even when very dilute. It is possible at 0° to have an agar gel containing only 0·1 per cent. of agar and at the ordinary temperature 0·5 per cent. of agar gives quite a stiff gel. When a dilute agar gel is kept, small droplets of water are exuded at the surface—a phenomenon known as *syneresis* (see p. 204). A noteworthy feature of the agar gel is that it does not adhere to glass and when cast in glass moulds can very readily be removed without the necessity of warming. On the other hand, gelatin adheres to glass most tenaciously and the gels can be removed without rupture only with the utmost caution, even after melting the surface by plunging the glass vessel into hot water for some seconds.

Kruyt and de Jong (*Kolloidchem. Beih.*, 1928, **28**, 1) have made a detailed study of the agar sol, particularly of the effects produced by electrolytes. The sol is negatively charged. An investigation of the simultaneous influence of electrolytes and dehydrating media has given strong support to Kruyt's theory of the stability of lyophilic colloids (p. 162). When the charge on the agar particles is removed by electrolytes in water as the dispersion medium, the sol still remains stable, but when the dispersion medium is alcohol-water or acetone-water the sol is flocculated under these conditions. Alcohol or acetone has the greatest influence at medium concentrations, for which the viscosity of their mixtures with water is greatest. It would therefore seem that the viscosity of agar sols is due not only to the charge on the particles, but also to hydration of the particles.

Soaps

By soaps are understood the salts of the higher members of the fatty acids. In practice, the alkali salts are of prime importance because of their cleansing properties, and most investigations have been conducted with the sodium and potassium salts of stearic, palmitic, and oleic acids. These are the constituents of ordinary commercial soaps.

The soaps occupy an interesting position in colloid chemistry. They dissolve in alcohol to form " true " solutions, which produce a normal rise of the boiling-point, but in water they dissolve in part as colloid and in part as true solution ; with increasing concentration of the soap solution the proportion of colloidally dispersed material increases, and the proportion of colloid also increases with the molecular weight of the acid. The greater part of our knowledge of soap solutions is due to the researches of McBain (*3rd Report on Colloid Chemistry, British Assn.*, 1920).

A study of the boiling-points of soap solutions (actually dew-point determinations are more serviceable) leads to the view that no electrolytic dissociation of the soap has occurred, but it is also found that soap solutions have a high electrical conductivity, which indicates considerable dissociation. This apparent contradiction could be explained, and was explained at first on the assumption that soaps were hydrolysed considerably in water and that the hydroxyl ions produced thereby were responsible for both the electrical conductivity and the rise of boiling-point, the free fatty acid or acid soap being osmotically inactive.

McBain's work showed quite clearly, however, that the concentration of free hydroxyl ions in the solutions is quite small, and that the discrepancy between osmotic and conductivity relations can be explained only on the assumption that electrolytic dissociation and colloid formation occur simultaneously. In this way, the increase in the number of particles due to electrolytic dissociation is balanced by the decrease due to colloid formation. The colloid is considered to exist in two forms: the *neutral colloid*, consisting of undissociated soap, and the *ionic micelle*, which is an aggregate of acid anions.

The sols can solidify in two ways, to form either a transparent gel or an opaque curd. For example, 0·6 N-sodium oleate solution when slowly cooled to 6° forms a transparent gel, which changes to a curd when kept for a day or two. The curd is produced at once if the cooling is rapid. The sol-gel transformation involves no change in electrical conductivity, but an increase in conductivity is observed when the sol sets to a curd. Ultramicroscopical examination of the curd reveals fibrillar and needle-like particles, but these cannot be observed in either the sol or the gel.

The detergent effect of soap was formerly attributed to its alkali content, which was supposed to react with the fatty materials in the clothes or skin being washed. Some experiments of Spring (*Kolloid-Z.*, 1910, **6**, 11, 109, 164) cast a new light on the matter. Soot was freed from fat by washing with alcohol, ether, and benzene. The soot was then found to form a stable suspension in water, which was readily retained by filter paper. When the suspension contained a small quantity of soap it passed through the filter paper without even staining it. It follows that the soot is not really filtered out by the paper, but is adsorbed by it; indeed, when the filter containing the soot was turned inside out and water was passed through, the soot remained on the paper; but soap solution immediately removed it. Adsorption and peptization appear to be the processes involved. The soap accumulates on the surface of the particles which normally adhere to the surface in question and destroys the adherence; it simultaneously gives an electric charge to the dirt particles, transferring them from the object to be cleaned to the soap solution.

Silicic Acid. The silicic acid sol is of particular interest from many points of view It is interesting historically, as it was known to Berzelius and formed the subject of many of the classical investigations of Graham; it has a geological interest, because gels of silicic acid are sometimes found during mining operations and there is reason to believe that many stones, such as the opal and agate, are dehydrated silicic acid gels, sometimes coloured with metallic oxides; and it has a special colloid chemical interest, since, although usually regarded as a hydrophilic colloid, it also exhibits some resemblances to the hydrophobic colloids; further, unlike the majority of hydrophilic colloids, it is inorganic and has a relatively simple chemical composition.

The sol can easily be made from sodium silicate. Commercial water glass is diluted with freshly boiled and cooled distilled water until its specific gravity is 1·6. A mixture of 30 c.c. of concentrated hydrochloric acid (d 1·2) and 100 c.c. of water is put into a beaker and 75 c.c. of the diluted water-glass solution are poured into the acid, during constant stirring. If the acid is poured into the sodium silicate solution a gel is formed instead of a sol. The resulting sol is dialysed to remove the sodium chloride, testing the dialysate with silver nitrate solution, but if dialysed too far the silicic acid may set to a gel. The colloid is highly disperse and in the initial stages of dialysis a certain proportion diffuses through the membrane. As much as 10 per cent. of the silicic acid may be lost in this way.

The dialysed sol is clear and colourless, gives little Tyndall cone and exhibits scarcely any inhomogeneity in the ultramicroscope. The highest concentration obtained is about 10 per cent., but such sols are unstable and on being kept set to a gel fairly rapidly. More dilute sols containing about 1 per cent. of silicic acid can be kept for years and are fairly stable towards electrolytes on the whole, but the addition of ammonia, phosphates, carbonates, and even carbon dioxide causes gel formation. The viscosity of the sols is not much greater than that of water, but increases slowly on keeping or on the addition of certain electrolytes, the sol becoming opalescent and finally setting to a gel, whereupon the viscosity rises very rapidly. The osmotic pressure decreases as the viscosity rises.

The gel formed from silicic acid differs from those of gelatin and of agar in at least two important particulars. It is almost non-elastic and it is irreversible in the sense that it cannot be re-transformed into the sol state by warming in water. To regain the sol it is necessary to treat the gel with sodium hydroxide and then repeat the process of adding hydrochloric acid and dialysing.

The particles are negatively charged in alkaline, neutral, and weakly acid solutions, but are positively charged in more strongly acid solutions, a variation of charge with hydrogen-ion concentration which is characteristic of hydrophilic colloids. The isoelectric point is thus on the acid side and it is found that the sols have their maximum of stability at the isoelectric point. This fact is contrary to what is observed in the case of hydrophobic colloids and also the majority of hydrophilic colloids. It must be remembered, however, that in sols of this type the stability probably depends not so much on electric charge as on hydration.

Colloidal silicic acid has no protective action on colloidal gold.

CHAPTER XIV
SOME NON-AQUEOUS COLLOIDS

Organosols. Colloids in dispersion media other than water have not so far received much theoretical study, but they are nevertheless important technically. On the whole, their examination reveals a closer similarity to aqueous colloids than might be expected. There are lyophobic and lyophilic sols in organic dispersion media just as there are hydrophobic and hydrophilic sols; the particles show Brownian movement and in numerous cases have been proved to carry an electric charge, although, as the dispersion medium is generally a good insulator, a sol often contains both positively and negatively charged particles; the lyophilic sols form jellies which swell in the dispersion medium, and may have a very high viscosity, and also can act as protective colloids to the lyophobic sols. The chief difference is in matters connected with ions, such as precipitation and peptization by electrolytes, and it is well not to lose sight of the fact that processes usually associated with and explained in terms of ions can occur in non-dissociating media.

According to Fischer and Hooker (*Kolloid-Z.*, 1930, **51**, 39), the importance of electrical phenomena in determining the stability and other properties of colloids has been over-estimated and emphasis is laid on the solvation of the colloid particles. The solvated type of colloid is described as having a considerable solubility for its own dispersion medium and differs in this respect from the suspensoid type. Soaps of aluminium and magnesium were found to give clear, stable gels in benzene, toluene, xylene, chloroform, carbon tetrachloride, ethylene dichloride, hexane, heptane, and decane. Cadmium stearate in toluene gave a clear solution at 100°, forming a dense white gel on cooling to 60°, which crystallized with exudation of a liquid phase at 6°. These authors conclude that the gel is a very viscous liquid-liquid system and that on further cooling it changes to a less viscous solid-liquid system.

Some further examples of recent investigations on non-aqueous colloids will serve to give a general idea of their properties. Among the suspensoid sols, reference has already been made to Svedberg's adaptation of Bredig's electrical disintegration method for the preparation of organosols of metals. Haurowitz (*Kolloid-Z.*, 1926, **40**, 139) has also described a method, using a high-frequency alternating current and a low voltage, which has been applied to the production of sols of iron, nickel, aluminium, lead, tin, zinc, copper, and magnesium in benzene. An alloy (brass) was obtained in colloidal solution for the first time. The sols can be stabilized by the addition of caoutchouc and will then keep for months.

An ingenious method for preparing such sols has been described by Roginski and Schalnikov (*Kolloid-Z.*, 1927, **43**, 67), and is based on the following principle. If a metal is distilled in a vacuum and the vapour is allowed to condense on a surface cooled by liquid air, the condensate is almost entirely amorphous, consisting of a highly disperse system of aggregated particles. By distilling simultaneously in this way a metal and a substance which is liquid at the ordinary temperature, a condensate is obtained, which, when melted, produces a highly dispersed sol of the metal in the liquid employed. The relative amounts of the substances in the condensate can be varied by suitable control of their distillation temperatures. The method has been employed successfully not only for many hydrosols but also for the preparation of organosols of mercury, cadmium, sodium, potassium, rubidium, and cæsium in dispersion media of benzene, hexane, xylene, toluene, and in some cases ether and alcohol. The method is particularly suitable for the production of sols of the alkali metals, since they are easily volatile and no carbonization of the organic medium occurs.

Questions relating to cataphoresis and other phenomena relating to electric charge in these sols are still somewhat obscure and probably vary considerably with the nature of the two phases. Whilst there are undoubtedly charges of considerable magnitude in many metal organosols, Prausnitz (*Kolloid-Z.*, 1930, **51**, 359) found that a fine suspension of glass particles in petroleum exhibited cataphoresis only under an applied potential of several thousand volts. In some cases it seems that the presence of small quantities of water has a great effect, for experiments carried out by Humphry and Jane (*Trans. Faraday Soc.*, 1926, **22**, 420) on the cataphoresis of a rubber sol in benzene showed that, although an undried sol exhibits charges of both signs, no appreciable charges are present when thoroughly dried materials are employed. Rubber sols are notably viscous and there appears to be a relation between the viscosity of such sols and the elastic properties of the rubber.

Sols of cellulose acetate and nitrate in organic media are of the lyophilic type and have been the subject of several investigations. The benzyl alcohol sol of cellulose acetate resembles the gelatin hydrosol in many ways, the substance swelling in the cold alcohol and dispersing therein at about 40° to form a slightly opalescent sol, which on cooling sets to a gel resembling the gelatin hydrogel in appearance and elasticity. The interesting sol-gel transformation of cellulose nitrate in amyl alcohol and benzene at elevated temperatures is described on p. 197.

When the organosols are treated with a liquid which is soluble in the dispersion medium, but does not dissolve the disperse phase, the latter is precipitated, presumably by withdrawal of the dispersion medium from the solvated disperse phase. This again illustrates the importance of solvation in the stability of colloids. According to experiments by Whitby and Gallay (*Trans. Roy. Soc., Canada*, 1929 [iii], **23**, III, 1), the case of precipitation of organosols of cellulose acetate by the addition of a non-solvent is affected by temperature to a much greater extent than is the relative viscosity, and the higher the concentration (and therefore the smaller the proportion of unbound solvent) the smaller is the volume of precipitant required to produce separation. It is concluded that precipitation by a non-swelling agent is not a trustworthy method of comparing the degree of solvation at different temperatures.

Camphorylphenylthiosemicarbazide gel has formed the subject of a recent study by Hatschek (*Kolloid-Z.*, 1930, **51**, 44). The substance dissolves readily at the ordinary temperature in hexane, benzene, acetone, and benzyl alcohol, the solubility rising with increasing temperature. On cooling, the substance crystallizes without formation of a gel. Gels are formed when hot solutions of the substance in toluene or in carbon tetrachloride are allowed to cool, whether rapidly or slowly. Gels containing up to 2 per cent. of the compound are clear, but at higher concentrations the excess crystallizes in the gel, and such gels in contact with the crystalline phase are stable for months. When solutions in benzene, toluene, or carbon tetrachloride are diluted with a non-solvent hydrocarbon (such as petroleum) unstable gels are formed, which eventually break up into macroscopic crystals. Gel formation is still observable at a concentration of 0·2 per cent.

Mercury as a Dispersion Medium. A quite new departure in the study of non-aqueous colloids has recently been begun by Rabinovitsch and Zyvotinski (*Kolloid-Z.*, 1930, 52, 31) using mercury as the dispersion medium. Iron amalgam was prepared by electrolysing a solution of ferrous sulphate with a mercury cathode. Such an amalgam separates into its component phases when kept, and the velocity of this separation was measured by placing the liquid amalgam in vertical tubes and at various intervals running off the lower portion of the contents and determining at what height in the tube a test for iron was obtained. From the velocity of rise of the iron in the amalgam the radius of the particles was calculated by application of Stokes' law. With increasing concentration the amalgam becomes more viscous and appears to approach the state of a gel; the eventual separation from this of mercury can be considered as a case of syneresis (p. 204). The velocity of separation of the phases is not appreciably affected by the addition of sodium amalgam or lead amalgam, but is increased by the addition of amalgams of zinc or tin. This behaviour is analogous to the influence of electrolytes on hydrosols. These observations indicate that iron amalgam must be considered as a colloid system. The system appears to be of considerable interest, because in other systems where the dispersion medium is liquid the stability is regarded as determined mainly by the electric double layer at the surface of the particles, and this is in turn connected with the dielectric properties of the dispersion medium. These properties are absent when mercury is the medium.

Pyrosols. Many metals dissolve in their fused salts to give cloudy solutions which are generally regarded as colloidal and are termed *pyrosols*. They are often formed in the electrolysis of fused salts and are responsible for some of the anomalous yields obtained in such cases. The simplest method of preparing pyrosols is to dissolve the metal directly in the fused chloride, e.g. lead in lead chloride, zinc in zinc chloride. Pyrosols are generally coloured and the disperse phase may be precipitated by the addition of sodium or potassium chloride. Coloured glasses are cooled pyrosols.

It appears from recent work by Heymann and his collaborators (*Z. physikal. Chem.*, 1930, **148**, 177 ; *Kolloid-Z.*, 1930, **52**, 269) that in some of the metallic " pyrosols " the metal is molecularly dispersed, although solvated, and that these systems are true solutions.

CHAPTER XV
THE PROPERTIES OF GELS

Setting. When a gelatin sol is cooled it sets to a gel or jelly, a process familiar enough in the preparation of table jellies and which goes under the term *gelatinization*, or, more broadly, *gelation*. Gelation is a characteristic of the hydrophilic colloids and is generally brought about by cooling or by evaporating the sol, whereas when a hydrophobic sol is evaporated the residue is generally a fine powder. Gelation may also be brought about by the addition of electrolytes, notably in the case of silicic acid, suggesting an analogy with coagulation rather than with crystallization. In the intermediate region there are the *gelatinous precipitates*, such as aluminium and ferric hydroxides. These are produced in precipitation reactions through the coagulation of the primarily formed hydroxide sol and are to be regarded as gels although they do not form one consistent whole like a gelatin jelly. Some prefer to call such gels produced by coagulation *coagels* and to refer to gels of the gelatin type as *jellies*. It is very likely, however, that the difference is a secondary matter, coagels consisting of particles of jelly separated by relatively large tracts of the liquid phase, and most authors prefer to make general use of the term " gel." It would appear that when gelatin occurs simultaneously throughout the whole liquid a uniform jelly is produced, but that a coagel is formed when the course of gelation follows certain surfaces. It is possible that in any case uniform gelation occurs initially, but is in some cases followed immediately by a process analogous to syneresis (p. 204), involving the separation of the dispersion medium. The present discussion refers to the setting of a whole sol, both dispersion medium and disperse phase, to a jelly.

As usual, it is difficult to define what is meant by a jelly, particularly as this form of matter has both solid and liquid properties (again depending on the inexact definitions of solids and liquids), but they may roughly be described as transparent or opalescent masses, readily suffering mechanical deformation, though offering some resistance, and having a fine structure which is less than microscopic. Most gels are decidedly elastic, whilst others have little elasticity.

Viewed from the standpoint of general colloid behaviour gelation suggests a coagulation process which takes place without separation of the dispersion medium. Actually, as will be described later, in many cases the setting of a gel is followed more or less slowly by a separation of the dispersion medium—a process known as syneresis—and the general impression received from all the observations is that the gelation of a hydrosol consists in the coagulation of heavily hydrated particles, enmeshing in the process the whole of the dispersion medium. In very dilute gels this cannot happen and gelation does not occur, whilst in concentrated gels there is less excess of dispersion medium to be retained by capillary forces between the particles, and such gels have properties more in accordance with those of solids.

So far as ultra-microscopical examination of the process of gelation goes, it offers support for this view. Bachmann (*Z. anorg. Chem.*, 1912, **73**, 125) observed the cooling of a 2 per cent. gelatin sol in the ultra-microscope; the initially diffuse light cone gradually developed into bright submicrons in vigorous Brownian movement, the motion becoming more sluggish as the size of the particles increased until it became merely a slight vibration, and the final picture was a mass of cohering particles without any motion. These observations seem to show that in gelating the primary particles unite to form secondary larger particles, which cohere and entrap the dispersion medium.

In any case it is not possible to accept the idea that during setting the hydration of the particles increases enormously so that there remains no free water to form a separate phase. Nevertheless, it seems quite certain that the particles of hydrophilic sols are heavily hydrated and that only a portion of the total water present is available to act as dispersion medium.

Experiments by de Jong (*Z. physikal. Chem.*, 1927, **130**, 205) indicate that the gelation of agar sols has all the characteristics of flocculation, but that in spite of this the electric charge and hydration of the particles remain practically unchanged. He has developed the hypothesis that charge and hydration are not uniformly distributed over the surface of each particle, but are localized in strongly protected spots, leaving other unprotected or hydrophobic spots on the particle surface. Coagulation of the particles takes place at these unprotected spots, leaving the charge and water of hydration free to line the inner sides of the spaces formed in the resulting spongy aggregates. It is suggested that when the temperature of a hydrophilic sol is lowered the number of hydrophobic spots on the particles increases without affecting the degree of hydration of the particle as a whole.

A feature of gelation is that the temperature of setting and of melting of the gel are generally different and sometimes to a remarkable degree. A 4 per cent. gelatin sol sets at about 28°, but does not melt below 31°; the hysteresis in the case of agar is, however, far more striking, gelation occurring at about 40° and melting at about 85°. Both the setting and melting temperatures increase somewhat with the concentration of the gel. They are not sharp temperatures in the way of melting points, but occur gradually over a small temperature range. The melting and setting temperatures of gelatin approach each other as the time allowed for the operation is increased. Banerji and Ghosh (*J. Indian Chem Soc.*, 1930, **7**, 923) have found that the melting and setting temperatures are identical, provided that sufficient time is allowed. The greater the concentration of the gel, the less is the time at which the melting and setting temperatures become the same.

The most characteristic feature of gelation is the enormous increase of viscosity, and naturally this has formed the most important study of the sol-gel transformation. The experimental curves are not characterized by a sudden break, but lead to the view that the sol-gel transformation is, unlike crystallization, a perfectly continuous process. There are several objections, however, to such an interpretation of the results. In the first place, the normal increase of viscosity with fall of temperature is great and may conceivably mask the effect which is due to gelation; moreover in addition to pure liquid viscosity, elastic properties gradually come into play and may obscure the effect.

There are, indeed, other data which indicate a discontinuity in the sol-gel transformation. The viscosity of a 10 per cent. sol of cellulose nitrate in amyl alcohol and benzene decreases with rise of temperature, but the sol sets to a gel when the temperature is *raised to* 70°. This unusual phenomenon is reversible, the gel melting to a sol once more on cooling. In this case gelation is quite distinct from the temperature variation of viscosity. A further interesting point has recently been described by Hatschek (*Kolloid-Z.*, 1929, **49**, 244) in a study of the formation of gas bubbles in gelatin gels. Although gas bubbles are spherical in viscous liquids, they are found to be lenticular in gelatin gels. The bubbles can conveniently be formed by diffusing acetic acid into gelation gels containing sodium carbonate. Observations under the microscope made during the setting of molten gels show that the spherical bubbles contained in the sol become lenticular quite suddenly. This indicates a definite discontinuity at some stage of the transformation.

The influence of neutral salts on the setting point is considerable. As a rule, sulphates and acetates raise the setting point, whilst nitrates, iodides, and thiocyanates lower it. Those substances which lower the setting point generally decrease the viscosity and aid dispersion, and vice versa.

Thixotropy. Thixotropy is a peculiar " unsetting " of a gel due to mechanical deformation and has been studied in recent years by Freundlich and his collaborators (*Kolloid-Z.*, 1928, **42**, 289). It is particularly noticeable with a concentrated ferric hydroxide sol which has been transformed to a gel by adding an electrolyte such as sodium chloride. When such a gel is vigorously shaken it liquefies to form a sol, but, when kept, the sol gelates once more. The process can be repeated. The phenomenon suggests that the properties of the gel are dependent on the preservation of a particular structure or mode of cohesion of the particles and that this structure is destroyed by the agitation, the normal coagulation being restored after a period of rest. It is well understood in the light of the theory of gelation given by de Jong (p. 196).

Even gelatin gels show the phenomenon of thixotropy, although in this case it cannot be observed easily owing to the great velocity of the reverse transformation. However, it was found that the cataphoretic migration velocity of microscopic particles of zinc and quartz in gelatin gels was practically the same as that in gelatin sols. Considering the enormous difference in the viscosity of the two media it can only be supposed that the electrical migration of the particles brings about a thixotropic loosening of

PROPERTIES OF GELS

the gelatin gel and that the particles really move through a narrow channel of sol, which becomes closed after their passage.

Swelling. It is characteristic of hydrophilic gels that when placed in water they absorb a certain amount and may increase in volume very considerably. Such an increase in volume of the gel means also that a considerable pressure must be set up. The process is known as swelling or imbibition and is a matter of importance and interest in everyday life. It is well known that wood swells when left in water or even in damp air and will cause such articles as windows and drawers to cease to fit. Cooking offers many excellent examples of swelling, and one cannot fail to be struck by the large rice pudding which is made from a mere handful of the dry rice.

For fundamental scientific investigation it is more suitable to investigate simpler substances such as a gelatin gel. An impression of the swelling of gelatin is gained from some early work by Schroeder (*Z. physikal Chem.*, 1903, **45**, 109). He found that a dry gelatin plate weighing 0·904 grm. when placed in an atmosphere saturated with water vapour at the ordinary temperature took up 0·37 grm. of water in eight days, after which no further gain occurred. When the plate was immersed in liquid water at the ordinary temperature it took up a further 5·63 grm. of water during the first hour and still more at a slower rate when left for a longer time. The amount of water taken up is therefore very great and there is a proportionately large increase in volume.

When removed from the liquid the swollen gel loses water readily, in the first stages as readily as from a pure water surface, and at the same time the gel shrinks. The loss occurs from the surface and is controlled mainly by the rate of diffusion of water from the interior. In a cube the loss is greatest at the corners and edges, and as these become tough and shrink as drying proceeds the piece of gelatin becomes very distorted. Hydration and dehydration in these gels are quite continuous processes and are thus distinct from the taking up of water of crystallization, where the vapour pressure curves show definite breaks.

Swelling is most conveniently studied by following the increase of volume or weight of the gel. A method frequently adopted is to use the gel in powder form and note the height to which a known mass of the powder swells when put in a test-tube containing the swelling medium. With substances such as gelatin and agar, which form gels that can be easily handled, plates or discs of the material may be made and the course of swelling followed by removing and weighing the swollen gel at various intervals of time.

The pressure produced in the swelling is very high. Swelling pressure is greatly involved in the growth of plants, and it is a fact that heavy stone slabs can be moved by the pressure exerted in this growth. It is sometimes said that a paving-stone can be moved by a blade of grass. An old method of splitting tree-trunks was to place dried peas in a cavity in the trunk and then pour water over them. The pressure exerted by swelling gels has been measured by placing the gel in a cylinder provided with a perforated piston so that the swelling liquid can have free access to the gel. The piston can be weighted and when the externally applied pressure is equal to the swelling pressure no swelling can occur.

Some values obtained by Reinke with such an instrument (called by him an " œdometer ") and working with dry discs of seaweed are given in the following table and are instructive in showing the large forces involved.

Applied pressure (atm)	Percentage volume increase.	Applied pressure (atm)	Percentage volume increase.
41 2	16	7 2	97
31·2	23	3 2	205
21 2	35	1 2	318
11 2	89	1·0	330

At atmospheric pressure the seaweed expands to 3·3 times the initial volume and even under a pressure of 41·2 atmospheres it still expands 16 per cent. Conversely, the figures indicate that a very large pressure is necessary in order to squeeze out the swelling liquid from a gel, and this is a point of importance in considering the structure of gels.

The swelling pressure falls as swelling proceeds, so that the greatest energy changes are involved in the first small quantities of swelling liquid imbibed, and consequently it becomes more and more difficult to remove the last traces of liquid. This aspect was studied by Posnjak (*Kolloidchem. Beih.*, 1912, **3**, 417), who used an apparatus similar in principle to that described above, the external pressure being provided by means of a gas cylinder. The swelling of rubber in a number of organic liquids was investigated and the curves connecting swelling pressure with the weight of liquid taken up show the enormous pressures obtained at the start of the process. The curves for different organic liquids are all practically parallel, showing that there is nothing specific in the behaviour of these liquids in regard to swelling.

A most important point to be observed is that although the volume of the gel increases during swelling the total volume of the system gel + liquid *decreases*. This can readily be verified by placing leaf gelatin in a small flask fitted with a stopper carrying a graduated glass tube, filling with water until it enters the graduated tube, and placing in a thermostat. The level of the meniscus will be observed to fall during the swelling. This means that either the water or the gelatin micelles decrease in volume during imbibition. Since the lyotropic series of anions is found to affect the process greatly it is most likely that a change in the sense of compression of the water molecules takes place (see p. 164).

The general impression obtained from a review of all the published work on the swelling of gels is that the process is a reversible solvation, or, so far as hydrogels are concerned, a continuous hydration of the particles. Each particle acquires a growing sheath of water molecules, so that the volume of the disperse phase increases, but possibly an orientation of the adsorbed water molecules in the sheath leads to closer packing than the random arrangement of molecules in liquids in the ordinary way, thus causing a decrease in the total volume of the system. Considerably more work needs to be done, however, before we can afford to accept a very definite view of these processes.

The compression occurring during swelling entails that heat must be liberated during the process, and calorimetric measurements of the process have shown this to be the case. The following table gives an idea of the changes involved, the heat of swelling being given in grm. cals. per grm. of gel.

Gel.	Heat of swelling	Gel	Heat of swelling.
Gelatin	5·7	Gum arabic	9·0
Starch	6·6	Gum tragacanth	10·3

The effect of aqueous solutions of electrolytes on the swelling of hydrogels is very marked. Most of the experiments have been carried out with gelatin. In general, the swelling is at a minimum at the isoelectric point and addition of either hydrogen- or hydroxyl- ions favours the process. Neutral salts also have strongly marked effects and the lyotropic series of anions is in evidence. In solutions of citrates, tartrates, or sulphates gelatin swells less than in water, but in solutions of iodides and thiocyanates the swelling is increased, and in more concentrated solutions both gelatin and other gels readily disperse even in the cold. Strong swelling generally leads to dispersion.

The dispersion of "refractory" gels by this means has received considerable study at the hands of von Weimarn (*Kolloid-Z.*, 1926, **40**, 120 ; 1927, **41**, 148 ; *Rep. Imp. Ind. Res. Inst. Osaka*, 1927, **8**, 67) with some very interesting results. Cellulose, fibroin, chitin, fibrin, and keratin become plastic and finally go into colloidal solution, when treated with concentrated aqueous solutions of very soluble inorganic salts. In order of influence these are $LiSCN > LiI > LiBr > LiCl$; $NaSCN > NaI$; $Ca(SCN)_2 > CaI_2 > CaBr_2 > CaCl_2$. In some cases it is necessary to heat the solution, but the rate of dispersion depends also on the purity, age, and previous history of the substance. The substances are reprecipitated by the addition of alcohol.

Silk dissolves at ordinary pressures to form a colloidal solution in hot, concentrated solutions of many very soluble salts of the alkali and alkaline earth metals. The order of influence of the anions is as follow : fluoride $<$ sulphate $<$ citrate $<$ tartrate $<$ acetate $<$ chloride $<$ nitrate $<$ bromide $<$ iodide $<$ thiocyanate ; the Hofmeister series thus being represented. The addition of concentrated solutions of the fluorides of ammonium or potassium, or the sulphates of sodium, potassium, ammonium, or lithium to a solution of silk in hot, concentrated calcium chloride causes aggregation and reprecipitation of the silk. The regenerated silk generally has a thread-like structure. Simple cooling does not cause gelation, but only an increase in viscosity. The silk is also regenerated by the addition of alcohol, or an aqueous or alcoholic solution of tannin. The silk reprecipitated by alcohol may be obtained as a solid block.

Microscopical and ultra-microscopical observations made with threads of natural silk in hot, concentrated sodium iodide solution showed that as the threads swell the ultramicro-fibrils of which they are composed gradually became curled in spirals. When the thread has the consistency of a viscous syrup, several of these spiral curls coalesce into an oblong drop. In the reverse process, threads drawn out of a silk coagulum, precipitated by a neutral concentrated solution of sodium citrate, ultimately display properties identical with those of natural silk threads, including the appearance of interference colours. In the state in which the curled fibrils are evident, the silk coagulum displays elastic caoutchouc-like properties. It is therefore considered that the elastic properties of caoutchouc may be due to the presence of similarly curled fibrils in an interfibrillar, viscous, or plastic dispersion medium, which permits the fibrils to become more or less straightened on stretching.

The phenomena of swelling and shrinkage of gels are of great importance biologically, especially in connexion with muscular contraction.

In some instances there is an apparent contradiction to the fact that colloidal fibres swell in a damp atmosphere and shrink in dry air. For example, violinists know that their E string is liable to snap when damp, indicating a contraction rather than an expansion on taking up water. Even an ordinary rope becomes more taut on a wet day. The reason is that both the catgut and the rope consist of spirally wound threads; the individual fibres become thicker on swelling, with the result that they tend to move away from the common central axis and in doing so must decrease the length of the whole string. It is concluded from recent X-ray spectrographic investigations that the ultimate units of muscle fibres have such a spiral configuration.

When tendons of rats' tails are placed in acid or alkali they swell and become shortened in length, although this does not occur in pure water. There is a limit to the contraction—viz. 33 per cent.—but the amount of contraction in acid is a measure of the extent of swelling.

A further investigation of collagenic fibres has been made by Heringa and collaborators (*Proc. K. Akad. Wetensch. Amsterdam*, 1926, **29**, 1081, 1092). Dark-field observation of longitudinal and transverse sections of various tendons showed that they consist of collagenic bundles intertwined in a spiral form, and a similar condition was found in microscopic preparations of skin and loose connective-tissue. Further evidence was obtained from experiments on the swelling and desiccation of tendons and of isolated bundles of collagen fibres. Torsion was found to occur in both swelling and desiccation, but in opposite senses. The conclusions are strengthened by the results of an X-ray spectrographic investigation.

Syneresis. When a fairly concentrated sol of silicic acid sets, either spontaneously or through the addition of a small quantity of a suitable electrolyte, a uniform jelly is first produced, but in the course of an hour or so small droplets of water are to be seen on the surface and these gradually increase in size, the gel meanwhile shrinking. Simultaneously, the gel ceases to adhere to the sides of the container, around which a film of liquid appears. This exudation of liquid from the gel, with shrinkage of the gel, was observed by Graham, who gave the phenomenon the name *syneresis*.

The phenomenon probably occurs in all gels to some extent, though it is not always readily noticeable, or only between certain concentrations. Whilst silicic acid gels undergo syneresis at high concentrations, it is best observed in the case of gelatin or agar by using dilute gels. A 1 per cent. agar gel shows the exudation plainly after standing for a few hours, but in certain gels, such as viscose (cellulose xanthate) in organic media the phenomenon is even more striking, the gel contracting to 50 per cent. of its original size. The liquid exuded is not quite the pure dispersion medium, but contains a little of the disperse phase.

Some recent research by Ferguson and Applebey (*Trans. Faraday Soc.*, 1930, **26**, 642) brings out some interesting facts concerning the syneresis of silica gel. The volume of liquid exuded during syneresis is independent of the shape and free external surface of the gel, but is identical with the contraction of volume of the solid gel. As the total volume of the system does not change, the process is probably a squeezing-out of liquid enmeshed in the gel and not a decrease in the degree of hydration of the particles. Experiments on the time-course of syneresis of gels of constant hydrogen-ion concentration showed that syneresis does not begin immediately after setting, and when once started it follows an S-shaped autocatalytic curve. The velocity of syneresis and the total volume of liquid exuded *increase* with the concentration of the gel. This is most unusual, for it means that most water is produced from those gels which contain least. Syneresis starts almost immediately in alkaline

PROPERTIES OF GELS

gels and proceeds very rapidly, but the final volume of liquid exuded is less than in acid gels. These authors consider that syneresis is due to the mutual attraction of the micelles due to residual valencies and is therefore greater in concentrated than in dilute gels.

The facts certainly suggest that the water exuded is that bound by capillary forces between the structural units of the gel and not water of hydration, and it seems to the author that syneresis should not be described as the opposite process to swelling, as is done by many. The fact that the total volume of the system does not change during syneresis destroys the idea that it is swelling reversed.

It seems rather that syneresis should be regarded as a continuation of the process of setting or gelation. We can imagine the heavily hydrated particles of the colloid coagulating so that they touch at spots (hydrophobic spots according to de Jong's view) and form a jelly structure in which the liquid dispersion medium is retained in thin films between the particles by capillary forces. As the process of gelation proceeds, coagulation continues, the particles touch at more spots and tend to move closer together, thus producing a shrinkage of the gel and squeezing out some of the *capillary bound water*. On this view, syneresis is part of gelation, whereas, if it were a reversed swelling, *water of hydration* would be exuded and then a change of volume would be expected.

The process of syneresis is probably of great biological importance, particularly in the study of secretion by glands.

Liepatov (*Kolloid-Z.*, 1929, 47, 21) prepared a series of colloidal solutions in which the ratio of molecularly disperse to colloidally disperse phase was graded by adding various amounts of alcohol to a solution of geranin in water. Experiments on the hydration of the particles were conducted. Gelation of a sol is considered to take place when the particles of disperse phase come into such proximity that their mutually attractive forces come into operation. Syneresis is considered to be a continuation of this process, the particles growing together so as to squeeze out the liquid between them. The velocities of gelation and of syneresis increase with the concentration of the colloid.

Kuhn (*Kolloid-Z.*, 1928, 46, 299) has made a critical examination of a considerable amount of published work on syneresis, and the following general conclusions are reached. Certain gels (silicic acid, caoutchouc, etc.) show increased syneresis with rising concentration, whilst others (dyes, starch, agar, cellulose acetate, etc.) behave in the reverse way. There appears to be no general rule for the influence of temperature. The degree of syneresis varies with the dispersion medium and is sensitive to small amounts of addition agents ; in general, the syneresis is greatest when the gel is in the least stable state.

Other Properties. Gels are often described as either *elastic* or *rigid*. The elastic gels, such as gelatin, swell more in the dispersion medium than the rigid gels, like silicic acid, which gains or loses water with comparatively no volume change. The distinction is not good, however, for although gelatin is markedly elastic when compared with silica gel, a tube containing the latter emits a musical note when struck, showing that it has elastic properties.

Under moderate stress gelatin gels show perfect elasticity so long as the application of the stress is not prolonged. After a time the gel becomes permanently deformed. For gelatin gels the modulus of elasticity under tension is approximately proportional to the square of the concentration of the gel. The modulus is not at its maximum immediately after setting, but increases for a few hours afterwards. The modulus is also affected by the presence of dissolved electrolytes and the lyotropic series is again in evidence.

When gels are not under strain they are isotropic, that is to say, their properties are the same in every direction. Under strain they become anisotropic and as the optical properties vary in different directions they become birefringent and give the usual colours under the microscope in polarized light. Usually, owing to unequal rates of cooling, setting, swelling, etc., gels are found to be birefringent in certain parts. The unilateral deformation of the strains may produce an orientation of the particles which is responsible for the double refraction. That such an orientation does occur in some cases is certain from X-ray spectrographic investigations. Katz (*Kolloid-Z.*, 1925, **36**, 300 ; 1926, **39**, 180) has shown that caoutchouc gives an X-ray spectrogram showing only amorphous rings, but when stretched it gives in addition a line diagram corresponding with an assemblage of crystallites arranged with their axes in the direction of stretching. The relative intensity of the lines increases with the stretching until finally the whole of the caoutchouc appears to be in the "crystalline" form. Similar results were obtained with gelatin.

The Structure of Gels. There is still considerable divergence of opinion about the structure of gels, but several of the accepted views approximate closely to the same thing when studied seriously.

The first important experimental investigations were conducted by Bütschli (Untersuchungen über Strukturen, Leipzig, 1898), who from observations under the microscope concluded that many gels have a fine honeycomb micro-structure, which could be rendered visible by several methods. For example, a gelatin gel after hardening with formaldehyde exhibited such a structure on microscopical examination. The structure suggested resembles that of a foam, the walls of the cells being made of gelatin and being filled with water, which is thus dispersed in the form of droplets. The disperse phase and dispersion medium have therefore changed rôles, the solid phase having become the continuous medium and the liquid the discontinuous phase. The view would certainly account for the rigidity of gels and the difficulty of squeezing out the water, and the idea of a phase reversal has had the weighty support of Wo. Ostwald.

Pauli and others have raised objections to any conclusions reached from experiments on these lines on the ground that it is not permissible to suppose that the gel has remained unaltered by the treatment with formaldehyde or other agent, but that the microscopic structure is produced during this treatment. Indeed, Copisarow (*Kolloid-Z.*, 1928, **44**, 319) has shown that by the diffusion of various tanning agents into a gelatin gel under certain conditions a macroscopic honeycomb structure is formed.

Ultra-microscopic observations of the gelation of gelatin carried out by Zsigmondy and his co-workers have not been able to confirm the honeycomb theory and show that the ultimate structure of the gel is very much finer than that assumed by Bütschli. These researches indicate that in the gel the micelles grow until they are separated by extremely thin films of the dispersion medium. The sol-gel transformation according to this view consists of a transfer of water from the continuous to the discontinuous phase, but there is no phase reversal. In favour of the view is the fact that substances known to affect the water distribution (e.g. alcohol and very soluble salts) affect also the sol-gel transformation markedly. Further, the rate of diffusion of electrolytes in dilute gels is very little different from that in water. This is possible if there are open water channels throughout the structure, but seems unlikely if the crystalloid had to pass through innumerable layers of the honeycomb membrane. In more concentrated gels the velocity of diffusion of electrolytes is less than in water, probably because the channels of water between the micelles are then so thin as to offer considerable resistance to the movement of the dissolved molecules. An idea of the size of these channels or pores can be gained from measurements of the vapour pressure of the liquid contained in the gel and the value $3\mu\mu$ has been calculated for a silicic acid gel.

It has also been suggested that in gelation the particles arrange themselves into fibres or hairs which are intertwined. Such an arrangement readily accounts for elasticity and the facility with which salts diffuse and in certain cases (e.g. in soap curds) there is direct evidence of the existence of such fibres.

All these views appear to be bound into a consistent whole by a theory of von Weimarn (*Rep. Imp. Ind. Res. Inst. Osaka*, 1928, **9**, 1) which is the result of experimental work extending over many years and which has the further advantage of linking up gelation with precipitation and crystallization.

According to his view, jellies may be classified according to the degree of dispersion of their primary structural elements, giving four main groups—macrocrystalline, microcrystalline, ultramicrocrystalline, and sub-ultramicrocrystalline—or according to their mechanical properties—resin-like jellies, paste-jellies, soft elastic jellies, and solid jellies (glasses). A classification according to the concentration of the disperse phase is rejected as unsuitable, but a division into temperature-reversible and temperature-irreversible jellies is considered to be useful. Von Weimarn makes a distinction between network jellies, in which the whole mass sets uniformly, and coarse cellular jellies, in which gelation proceeds only on definite surfaces within the liquid. By thorough shaking of a coarse cellular jelly it is often possible to obtain a coherent network jelly, through breaking up of the membranes, although this is neither uniform nor transparent. The membranes of the cells of coarse cellular jellies are likened to plates cut from a network jelly of the same thickness.

The greatest experimental advances have come from the study

PROPERTIES OF GELS

of the formation of gels of the sulphates of calcium, strontium, and barium by precipitation when the solubility is made as small as possible and the potential degree of supersaturation as large as possible (cf. p. 9), and it is believed that there is no fundamental difference between these gels and ordinary jellies such as gelatin and agar. The conditions necessary for the formation of network jellies, cellular jellies, and mixtures of both kinds for the cases of the insoluble sulphates have been worked out and have then been applied to typical jellies of the gelatin type.

The production in a medium of aqueous alcohol of colloidal solutions of the calcium, strontium, and barium salts of sulphuric, chromic, tungstic, molybdic, hydrofluoric, oxalic, tartaric, citric, carbonic, boric, orthophosphoric, arsenic, arsenious, hydrosulphuric, and silicic acids and also of silver citrate, chloride, bromide, and iodide has been studied at temperatures down to $-30°$, and the following conclusions have been reached. (1) Where only one type of salt can be formed (e.g. barium sulphate, silver chloride) colloidal solutions are readily obtained, and excess of one component (barium ions for barium sulphate) gives a very stable colloid, whilst excess of the other component (sulphate ions for barium sulphate) diminishes the stability or causes coagulation. (2) Where more than one salt can be formed (e.g. carbonates, phosphates, silicates), instead of stable colloidal solutions there are formed voluminous gelatinous precipitates, or, under suitable conditions, network jellies.

The total amount of water is supposed to consist of (1) interatomic hydration, e.g. water in crystal hydrates (Al_2O_3, $3H_2O$; $SrSO_4$, $2H_2O$), (2) surface hydration, or the sheath of adsorbed water molecules surrounding each particle, and (3) structure hydration, or water enclosed by the pores of the structural elements of the jelly and held there by capillary forces.

The examination of a large number of photomicrographs of gels of hydrated strontium sulphate taken at different stages of their formation throws a most interesting light on the structure of these gels and lends colour to the view that the structure of gelatin and other jellies is of the same type, although their units may be too small to be rendered visible. The photographs demonstrate clearly that the tissue of the jelly is composed of bundles of thin crystalline needles ; in each bundle, the needles spread radially from one centre, and the ends of the needles of different bundles are interwoven, forming a continuous structure.

The following stages in the process of gelation can be observed : (1) formation of crystallization centres ; (2) development of these centres into spheres composed of radially disposed needles ; (3) union of the spherical bundles into flakes ; (4) filling of the entire volume by the flakes, with complete soaking up of the mother solution.

On keeping, these gels disintegrate, with the formation of a flocculent precipitate composed of aggregates of tangled needles, a process reminiscent of syneresis. The needles are considered as primary structural units of the jelly, and the spherical aggregates as secondary structural units. The gels are easily destroyed by energetic shaking, because the interlocking of the needles is disturbed, recalling the phenomenon of thixotropy.

From the study of a number of photomicrographs of precipitated gold and of some other substances it is concluded that gelatinous precipitates are particular cases of flocculent precipitates ; the individual grains of the former are ultramicro-crystals and the greater the degree of hydration the more the precipitate assumes the properties of a jelly. There is no evidence that the ultimate particles of a jelly have of themselves a gelatinous nature.

Von Weimarn concludes from all this work that the unit of all jellies is at least sub-ultramicrocrystalline, and that no true emulsoid jellies, consisting entirely of liquid phases, exist. Discussing the examination of jellies and glasses by means of X-rays he concludes that the individual structural elements are not devoid of an inner vectorial structure ; absolutely amorphous jellies and glasses do not exist. It is stated as a general law that any substance can under suitable conditions be obtained as a stable jelly.

The most important conceptions of gel structure are represented schematically in a rough way in Fig. 18.

Views closely similar to those of von Weimarn have recently been developed by Thomas and Sibi (*Rev. Gén. Colloïd.*, 1929, **7**, 295 ; 1930, **8**, 68, 105). They have conducted experiments on the crystallization of *l*-arabinosazone, which is soluble in hot water and readily crystallizes on cooling to form clusters of needles which interpenetrate to give a firm mass. When crystallization takes place in the presence of small amounts of gum, microscopical

PROPERTIES OF GELS

examination of the units of structure showed them to be intertwining hairs instead of straight needles, and it was found that the flexibility of the hairs as well as their length could be varied by altering the proportion of gum in the mixture. The experiments demonstrate that there is a relation between the flexibility and length of the hairs, their ability to intertwine so as to form a coherent mass, and their power to retain water, and that these

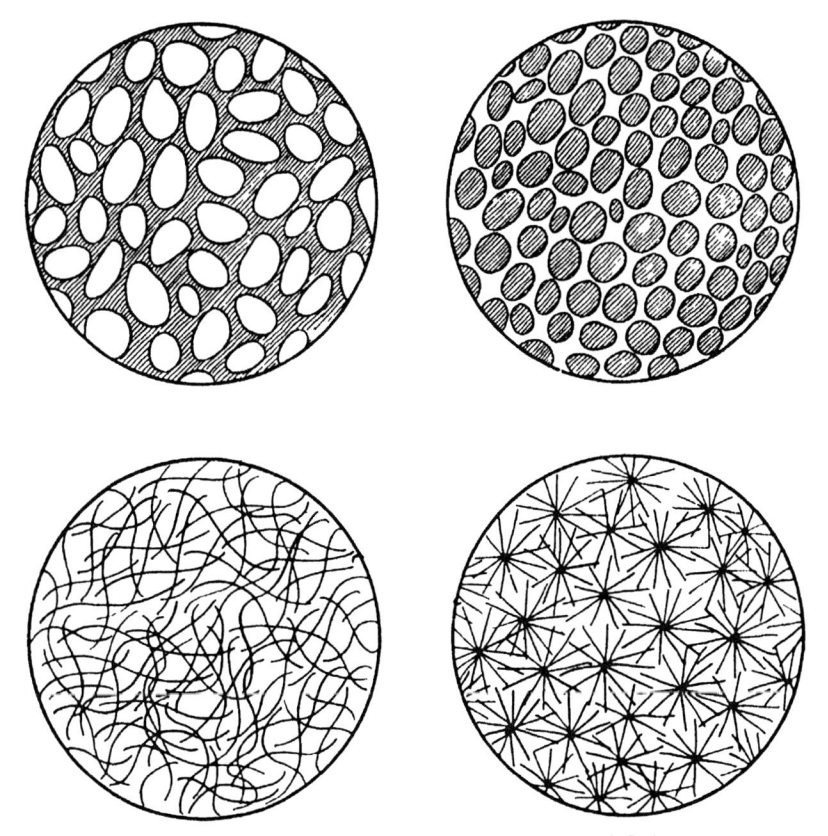

FIG 18 —Representations of the Structure of Gels

factors can be varied by the presence of impurities. When sorbitol acetate, benzoylcystine, benzoyltyrosine, quinine sulphate, optoquine sulphate, lithium urate, eucupine acetate, and apiine in aqueous solution are cooled they form transparent jellies, which initially appear homogeneous under the ultramicroscope, but when kept or on the addition of alcohol or acetone the jellies are transformed into masses of crystals. These are in the form of long, flexible, intertwining hairs, and when stirred or shaken the structure is destroyed. The separation of the dispersion medium is considered to offer an explanation of the general phenomenon of syneresis. The crystals may be filtered from the dispersion medium, dried, and redissolved, and when the sol is cooled the jelly is formed again. These observations support the view that jellies contain a solid phase and a similar fibrillar structure is suggested for typical jellies such as gelatin and agar. It is pointed out that gelatin and agar jellies are rendered more fluid by repeated mechanical rupture. An interesting feature appearing from this work is that gels of eucupine acetate having a concentration of only 0·025 per cent. were used and that under favourable conditions of hydrogen-ion concentration a gel of apiine having a concentration of only 0·07 per cent. can be obtained.

Thorne and Smith (*Kolloid-Z.*, 1929, **48**, 113) have examined the properties of calcium acetate gels with the object of obtaining a gelatinous substance of definite chemical composition. The gels were prepared by pouring a saturated aqueous solution of calcium acetate into alcohol. With larger amounts of calcium acetate, sols of the salt in water-alcohol mixtures were formed. Most of the gels formed in this way are not stable for more than twenty-four hours; they are opalescent at first and gradually soften with time, small nodules of calcium acetate eventually settling out. In gels containing a relatively large amount of water, needle-like crystals form radially from numerous crystallization nuclei, and it is suggested that gelation is an intermediate stage between true solution and crystal formation. The stability of the gels is increased, in some cases to six months, by the addition of acetone or various oleates, and in these gels a thread-like structure is visible to the unaided eye. Gels containing sodium oleate exhibit syneresis. When dried, the gels become turbid and alcohol is lost, the calcium acetate dissolving in the remaining water. Viscosity measurements indicated that the change from sol to gel takes place between sharp limits.

Wo. Ostwald (*Kolloid-Z.*, 1928, **46**, 248) has reviewed systems of classification of gels and has proposed a new scheme, which is based on the mode of production of the gel. Different types thus arise according to whether the gel is formed by falling of temperature, by chemical reaction, by coagulation, swelling, fermentation, or by geological influences. Allied systems include undercooled melts and solutions, slimes and foams. A distinction is also made between " lyogels," which are rich in liquid, and " xerogels," which are in a relatively dry state. Gels are defined as systems rich in liquid, of varying composition (liquid-liquid, liquid-solid, solid-liquid, etc.) and of varying degree of dispersion, though mainly colloidal, the particles of which either by mechanical growth or through other causes are in such proximity to each other that the liquid exists mainly in the form of lyospheres.

Wagner (*ibid.*, 1929, 47, 19) has suggested the terms " liquogel " and " viscogel " according to whether the sol formed on melting the gel has a high or low viscosity. Plastic or pasty substances are termed " plastogels " or " elastogels," the latter having higher elasticity.

The Solid-phase Rule. A discussion of gels is not complete without a description of some recent investigations of the equilibrium relations existing between gels and their solutions. The solubility of a simple salt in water is independent of the amount of the salt, as is well known, but this is not the case with a large number of colloids investigated. The solubility and other relations between the solid phase and the liquid dispersion medium vary with the amount of the solid phase, generally passing through a maximum for medium amounts. This relation has been called by Wo. Ostwald the " solid-phase rule " (Bodenkörperregel). The significance of the rule is not yet clear and its cause probably lies in the complex nature of the colloid-electrolyte complexes. With quite pure materials the rule may even not hold, but since the observations with ordinary materials are experimental facts they are of practical importance.

Wo. Ostwald (*Kolloid-Z.*, 1927, 41, 163) found that, contrary to the behaviour of molecularly dispersed substances, the amount of a colloidal gel dissolved or peptized by a medium is not independent of the amount of gel present, but generally increases with the amount of gel used. The first supposition was that the gel is heterogeneous in that it exhibits " state-isomerism," the " isomerides " varying in their susceptibility to peptization.

Von Buzágh (*ibid.*, 169 ; 1927, 43, 215, 220) showed that in he peptization of colloidal barium carbonate, casein, Congo-red acid, Congo-blue, and ferric hydroxide, the quantity of colloid which passes into solution is not independent of the amount of colloid taken. As a rule, the curve connecting these quantities exhibits a marked maximum, showing that for a given peptizing agent and a given volume there is an optimum condition for sol formation when a medium amount of colloid is present. A similar relation was found when a sol was used as a peptizing agent in place of an electrolyte, for example, in the peptization of charcoal and ferric hydroxide by soap solutions and of kaolin by humic acid.

Ostwald, Steinbach and Köhler (*ibid.*, 1927, **43**, 227) found that the solid-phase rule holds not only for colloids, but also for coarse suspensions and is applicable to the stabilization of suspensions of charcoal by picric acid, aniline, and pyridine. Köhler (*ibid.*, 1928, **45**, 345) was able to establish the rule where the disperse phase was a liquid. The amount of olive oil or pea-nut oil containing oleic acid which can be emulsified by dilute sodium carbonate or sodium hydroxide solution depends on the amount of oil, when the concentration of alkali and the total volume of the aqueous solution remain constant. The amount emulsified is greatest for a medium amount of oil, and the optimum amount of oil increases with increasing concentration of alkali. It is clear that observations such as these may have an important technical bearing.

Ostwald (*ibid.*, 1927, **43**, 249) later suggested an explanation in terms of adsorption. If the amount of solid phase is small, adsorption is great and the substance tends to be precipitated; on the other hand, with a large excess of solid phase adsorption is so small that little colloid-chemical effect is produced. Consequently, peptization will be greatest for medium amounts of the solid phase. Von Buzágh (*ibid.*, 1928, **46**, 178), however, concluded from a review of the work that other factors than adsorption are involved in the solid-phase rule and suggested that the variation of particle size of the suspension with the concentration also plays a part. Experiments conducted with suspensions of animal charcoal, calcium carbonate, barium sulphate, zinc oxide, red lead, and mercuric oxide in water indicated that the solid-phase rule holds in such coarsely disperse systems in the absence of an additional peptizing agent. The degree of peptization was found to depend, not only on specific adsorption, but also on the concentration of the suspension, these factors determining whether or not peptization in the narrow sense takes place.

Von Buzágh (*ibid.*, 1929, **48**, 33) also showed that the solid-phase rule holds for the charge on the particles. Measurements of the cataphoretic migration velocity were carried out with suspensions of animal charcoal in picric acid solutions and with suspensions of kaolin and of bentonite in solutions of sodium hydroxide, using different concentrations of the electrolyte as peptizer and varying amounts of the solid phase. In each case, the migration velocity for a constant amount of solid phase was found to rise at first with increasing concentration of the electrolyte, reach a maximum, and then fall. A similar curve with a single maximum was obtained when a constant concentration of peptizing agent was used and the migration velocity was plotted against the amount of solid phase. The similarity in the two curves is to be expected on the assumption that a definite electric charge on the particles is necessary for peptization and that this charge is related to a definite adsorbed amount of the peptizing agent, for with small amounts of solid phase the relative amount of adsorbed electrolyte is too high, whilst with large amounts of solid phase there is insufficient electrolyte to stabilize the particles. The maximal value of migration velocity was practically the same, whether obtained by varying the amount of the peptizing agent or of the solid phase. A maximum was also observed in the curve connecting the degree of peptization with the amount of solid phase, the position of the maximum corresponding with that of the maximum of electric charge.

Reviewing these publications, Ostwald and Rödiger (*ibid.*, 1929, **49**, 314, 412) have distinguished four types of peptization: (1) adsorption peptization (e.g. carbon and picric acid); (2) dissolution peptization (metal hydroxides and acids); (3) peptization through swelling (gelatin in water); (4) spontaneous colloidal dissolution (dyes and soaps in water, cellulose acetate in chloroform). In adsorption- and dissolution- peptization an optimum of peptization is observed for medium amounts of the solid phase, and in spontaneous colloidal dissolution and peptization through swelling the solubility curve rises steadily as a rule. Experiments on the peptization of stannic acid by hydrochloric acid, zinc hydroxide by sodium hydroxide, and bismuth hydroxide by lactose and by sucrose furnished results in accordance with the solid-phase rule, but exceptions appeared in the systems zinc hydroxide-hydrochloric acid and zinc hydroxide-acetic acid, where the solubility is independent of the amount of the solid phase. It is held that in the exceptional cases molecular chemical reaction is chiefly concerned, whilst the solid-phase rule applies where colloid-chemical processes are mainly involved.

Jermolenko (*ibid.*, 1929, **49**, 424) has discussed the factors determining colloid solubility in relation to those affecting molecular solubility. Experiments were directed to establishing the validity of the solid-phase rule for the dissolution of copper carbonate by ammonia in the presence of ammonium chloride. The relation between the amount peptized and the amount of solid phase shows that the process is an example of dissolution-peptization.

The solid-phase rule has been criticized by Sörensen and Sladek (*ibid.*, 1929, **49**, 16), but Ostwald (*ibid.*, 209) has replied that the cases quoted by them are complicated by the chemical formation of new phases.

The solid-phase rule has also been observed in the swelling of gels. Ostwald and Köhler (*ibid.*, 1927, **43**, 233) found that the swelling of gelatin is influenced by the ratio of the solid phase to the volume of liquid, the relative swelling falling as this ratio increases. Further, the spontaneous dissolution of gelatin in pure water increases with increasing amount of the solid phase. The observation that the influence of the amount of the solid phase on both the relative swelling and the spontaneous dissolution of gelatin is markedly less at lower temperatures has considerable significance. In a further paper by Ostwald and Kestenbaum (*Kolloidchem. Beih.*, 1929, **29**, 1) attention is directed to the fact that in most researches on swelling the solid-phase rule has been neglected and doubt is thus thrown on the validity of earlier measurements of swelling. In general, the specific swelling increases with the ratio of the amount of swelling liquid to the solid phase. The effect was observed in the swelling of agar and hide powder and some detailed experiments were carried out on the swelling of gelatin in water and electrolyte solutions. A theory of the solid phase relations was developed from experiments on the electrical conductivity and nitrogen-determination of the part of the gelatin going into solution on swelling, experiments on swelling in electrolytes, the swelling of electrolyte-free gelatin, and the osmosis of gelatin sols. These experiments lead to the general view that solid phase relations are due to soluble substances in the gel, which dissolve in the swelling water to give a solution the concentration of which is proportional to the amount of solid phase; the solution thus produced affects the course of the swelling. In the case of gelatin in particular, the swelling in pure water depends on the content of calcium sulphate and of degradation products of gelatin. Calcium sulphate hinders swelling and therefore swelling is greatest with large amounts of the swelling liquid, in which the calcium sulphate forms a dilute solution. The theory is well supported by experiments in which substances known to affect the swelling were added; in these experiments the curve representing the solid phase effect was strongly influenced.

Lloyd (*Kolloid-Z.*, 1929, **48**, 342; 1931, **54**, 46) states that the influence of the volume of liquid on the swelling of commercial samples of gelatin is due to alteration of the relative concentration of electrolytes and degradation products present as impurities. The observation that the degree of swelling of gelatin in sodium hydroxide solutions varies with the volume of solution is due to hydrolytic decomposition of the gelatin. When pure gelatin swells in sodium hydroxide solutions at a sufficiently low temperature (0°) to prevent hydrolysis, the amount of swelling is independent of the volume of liquid.

Thermodynamics of Colloids. The foregoing considerations raise the question whether the laws applicable to molecular solutions can be applied to colloid systems, and, if so, how far. Little progress has been made along these lines so far, but reference may be made to a recent attempt by Cofman (*Rev. Gén. Colloïd.*, 1929, **7**, 337, 406) to apply to colloidal systems the laws applicable to gases and molecular solutions by finding what properties are truly analogous to pressure, volume, and temperature. For the pressure factor the swelling pressure of gels is taken, the volume factor is the volume of the colloidal system minus the volume of the colloidal particles themselves (corresponding with the term $V - b$ in van der Waals' equation), and consideration of the variation of surface tension with hydrogen-ion concentration leads to the replacement of the temperature factor by the hydrogen-ion concentration. Measurements of the surface tension of solutions of sodium oleate at various hydrogen-ion concentrations show that the surface tension is a linear function of p_H. The surface tension is also proportional to the surface concentration. The spontaneous dispersion of colloids is compared with the boiling of liquids. These ideas are developed in an application of the Carnot cycle to colloidal systems, replacing the usual variables by the analogous factors, and the efficiency of "colloid machines" working between two limits of hydrogen-ion concentration has been considered in this light. Such a study may have some importance in relation to muscular action.

CHAPTER XVI
DIFFUSION AND CHEMICAL REACTION IN GELS

Diffusion. Soluble electrolytes and other compounds diffuse readily into gels which are immersed in their solutions, a process which is of practical importance in the developing and fixing of photographic films and the tanning of leather, apart from its interest in biological reactions and geological formations.

This ready diffusion of crystalloids was noticed in the early days of colloid investigation by Graham (*Liebigs Ann.*, 1862, **121**, 5, 29), who also observed that colloids do not diffuse through gels. He concluded from his experiments that such a solute as sodium chloride diffuses in a gelatin gel at practically the same velocity as in water. Later measurements by Bechhold and Zeigler (*Z. physikal. Chem.*, 1906, **56**, 105) have shown that this is correct so far as dilute gels are concerned, but that the rate of diffusion is lower in more concentrated gels. As shown in the last chapter, such a behaviour is quite in accordance with the present-day views of the structure of gels.

An important feature of physical and chemical changes in a gelatinous medium is the absence of turbulence. Diffusion in liquids is complicated by convection currents and mechanical disturbances, but in a gel the process can take place steadily. Nevertheless, in gels there are other complicating phenomena. Diffusion is generally studied by allowing the gel to set in the lower part of a test-tube and then pouring the diffusing solution over it. Some gels (agar, for example) do not adhere firmly to the walls of the tube and the diffusing liquid penetrates the narrow channel left more quickly than the body of the gel. There are means of overcoming this, but in any case it may be avoided in firm gels by immersing the gel without any casing directly in the liquid. A more serious matter is that different solutes affect the swelling of the gel to different extents, so that there is not only a transfer of solute, but also of solvent, and the variation in volume of the gel complicates matters. In addition, it is often not possible to regard the gel as a chemically inert system and if it binds a certain proportion of the diffusing agent by chemical bonds, the course of diffusion is necessarily altered. Adsorption and electrical changes may also play a part, with the result that specific effects appear to enter.

Qualitatively, the difference in the diffusion of crystalloids and colloids in gels can be shown by allowing sols of gelatin or agar to set in test-tubes, so that each tube is about half full, and then pouring a coloured solution into the upper part of the tube. Solutions such as copper sulphate or potassium dichromate can be seen to have diffused some way into the gel after several hours, but night-blue and other coloured sols do not penetrate the gel. However, it must be remembered that a colloid will be in equilibrium with a certain amount of molecularly dispersed material. The dispersion medium of a gold sol, for example, will normally be saturated with respect to gold molecules. Often, this " true " solubility of the colloid is exceedingly small, but where it is not too small and where the molecular weight of the dissolved portion is not too high, this portion may diffuse at an appreciable rate. As it is removed gradually by diffusion, more of the colloid will dissolve to replace it and eventually, as with some dyes, all the colloid may have diffused away in a more highly dispersed form. Thus, although colloids, as such, do not as a rule diffuse appreciably, they may in certain cases attain the same result through slow transformation into the molecularly dispersed form which is in equilibrium with them. Naturally, the diffusing solution is always very dilute and the process is therefore slow, but the so-called oligodynamic action of metals is probably due to this cause. A colloid can thus act as a constant reservoir of material for an exceedingly dilute solution. This fact may be capable of important practical developments in the future, for it offers a means of supplying continuously an agent at a low concentration. This cannot be done by the use of very dilute solutions in the ordinary way except by using an enormous bulk of the solution, because the small amount of solute present soon becomes used up. This principle appears to underlie the use of colloidal preparations which have of recent years been applied for medical purposes. As our bodies are mainly colloidal structures and most of our food is colloidal, there has been a suspicion that medicines are best administered in the colloidal state. Whilst there is very considerable dissension of opinion about this as a principle, there is no doubt that colloidal silver and many similar preparations are beneficial in certain cases. It appears likely, however, that the effect of these preparations is not due to the particles of colloid, but to a very dilute solution which is kept up to constant concentration by the colloid particles acting as a reservoir.

Quantitatively, although diffusion in gels might constitute a convenient mode of investigation of diffusion laws because of the fixity of the medium, not only is the problem beset with all the difficulties of interpretation which have been mentioned, but there are practical difficulties in the observation of the process. Even in the diffusion of a strongly coloured electrolyte such as potassium dichromate it is not possible to fix the exact boundary, for the concentration gradient falls gradually to zero and the boundary is consequently diffuse. Sometimes a perfectly sharp boundary is produced; for example, in the diffusion of hydrochloric acid into gelatin gels, where, although the diffusing agent is not coloured, it causes a notable change in the refractive index of the gel and clears up the usual turbidity. In this case, however, the swelling of the gel and chemical changes complicate the matter. Ferric chloride and chrome alum solutions also produce a sharp boundary of diffusion, but here there is considerable hydrolysis, the metallic hydroxide being retained by the gel, whilst the free acid wanders on quickly.

In order to produce a sharp diffusion boundary, the progression of which can be followed with ease, many investigators have adopted the expedient of adding a suitable indicator to the gel. For example, the course of diffusion of sodium chloride into gelatin can readily be followed by using a gel which contains a very dilute solution of silver nitrate, so that the white silver chloride formed shows how far the sodium chloride has penetrated. Such methods are open to objection, for, apart from the fact that the indicator itself diffuses in the opposite direction, some of the diffusing agent is disposed of in the chemical reaction and the precipitate produced may obstruct the diffusion.

Others have attempted the quantitative determination of the amount of diffusing agent in different parts of the gel. This is a difficult process and generally gives poor results, because when carried out by volumetric methods the end-point is usually very vague in the presence of the gel colloid, and gravimetric methods are equally complicated by the variable ash content of the gel, among other objectionable features. A chemically inert gel of definite chemical composition would be a real boon in investigations of this sort; practically all the work has been carried out with natural products of variable and unknown composition, such as gelatin and agar. Certain gels in organic media would probably fulfil some of the conditions, but as our knowledge of solutions and chemical phenomena therein is mainly confined to water as a solvent, results obtained in organic media have scarcely the same significance.

Another method of following the diffusion has been to remove the gel from the test-tube and perform spot-tests with a suitable indicator to detect how far the diffusing electrolyte has penetrated.

Among the many abnormalities and specific effects the investigations have revealed certain regularities. The diffusion of a soluble substance from an aqueous solution into a gel or from the gel into water follows the ordinary diffusion law of Fick. When the gel already contains one electrolyte and the superimposed solution contains another, both diffuse in opposite directions in accordance with Fick's diffusion law, independently of the concentrations and osmotic pressures of the two electrolytes, so long as they do not react to form an insoluble substance. Even if they do react, but incompletely, as in the case of magnesium chloride and ammonia, some of the electrolyte originally in the gel will be found in the upper liquid layer and some of that in the superimposed layer will be found in the gel.

When chemical reaction of the two electrolytes produces a precipitate which is not soluble in excess of either reagent, the matter is quite different and diffusion is a one-sided process so far as the phase-boundary at the gel-liquid surface is concerned. The solution of higher osmotic pressure diffuses into that of lower osmotic pressure. For example, if a fairly concentrated solution of silver nitrate is poured over a gelatin gel containing a dilute solution of sodium chloride all the silver chloride precipitate is formed in the gel and not in the solution above. At the same time, the sodium chloride does move towards the diffusing silver nitrate, because the precipitate of silver chloride rarely extends quite to the bottom of the tube, but leaves a dead-space. This apparent chemical attraction has been studied particularly by Liesegang. This one-sided feature of diffusion, which is by no means invariably valid, is generally known as Pringsheim's Rule (*Jahrb. wiss. Bot.*, 1895, **23**, 1).

Crystallization and Precipitation. The conditions determining the size of the particles produced in crystallization and precipitation were discussed in Chapter III, where it was shown that a given mass of matter may consist of a large number of small particles or of a small number of large particles according to whether the rate of formation of crystallization nuclei is greater or less than the rate of growth of the nuclear crystals.

These factors and others are modified by the presence of a gel, with the result that the forms produced may be quite different. Since convection and mechanical disturbances are ruled out in the gel, fresh solute can reach the growing crystals only by diffusion, and except in dilute gels this is a slower process than in water as a medium. Consequently, if this were the only factor operating it would be expected that extremely well-developed crystals would be produced, for the more slowly a crystal grows the more perfect is its form.

In practice this is often the case and particularly regularly formed crystals of large size are often grown in gels of silicic acid and also of gelatin and agar. Reference may be made to the large crystals of gold made by Hatschek and Simon (*Kolloid-Z.*, 1912, **10**, 265) by reducing gold chloride in a gel of silicic acid. In general, however, it is not possible to predict what influence the gel will have, because there are other factors which cannot always be anticipated. Whilst retarding the rate of growth of single crystals tends towards the production of large, well-formed specimens, provided the rate of formation of nuclei is small, the same cause would produce a mass of small crystals if the rate of formation of nuclei is great, because fresh nuclei will be forming while the existing nuclei are slowly growing. Again, the gel may also affect the formation of nuclei. This is a process which is not completely understood, but, if the formation of a nucleus depends on the probability of a group of molecules being arranged in a manner roughly conforming to the space-lattice, then it would be expected that the presence of a colloid would diminish the probability. In that case, large crystals would again be favoured. It is possible that the apparently higher solubility of certain compounds in gels than in water (compare Hedges, *J. Chem. Soc.*, 1929, 2779) may be ascribed to a supersaturation due to this cause. On the other hand, in certain cases the gel itself may be able to provide nuclei for the material. Another weighty factor having the reverse effect, i.e. producing a small particle size, is the protective effect of the gel. A gelatin gel may be expected to protect the small particles of a precipitate in the same way as it protects a gold sol, and for this reason precipitates formed in gelatin are often highly disperse, if not colloidal. This is not invariably the case, however, and diffusion reactions often produce large crystals even in gelatin. This consideration is well borne out by the fact that as a rule silicic acid gel, which has no perceptible protective effect, leads to the formation of much larger crystals than are formed in gelatin. The result depends, therefore, on which of two opposite effects is the stronger.

Another feature of precipitates or masses of small crystals in gels is the subsequent slow growth of the larger particles at the expense of the smaller. This so-called "ripening" is of importance in photographic films.

Crystallization in gels is further modified in that the crystal habit may be quite different. Sometimes the growth of certain faces is retarded more than that of others, but often there is a tendency to form spherulites. This tendency towards rounded contours is reflected in the arborescent forms of crystals often produced. Such forms generally indicate the action of surface forces and are characteristically produced when thin films of substances crystallize.

Chemical Reaction. Precipitation reactions between solutions are usually conducted in test-tubes to the accompaniment of shaking. For qualitative tests this clumsy method of causing the solutions to react suffices, but in many quantitative operations it is recognized that the resulting multitudinous variations of concentration of the solutions in different parts of the vessel, with the variations in grain size of the precipitate and in the kind and amount of adsorbed electrolytes, tend to produce irregular results and as far as possible uniform mixing is adopted, such as by dropping one solution into the other slowly with constant stirring.

From this point of view, precipitation reactions in gels are ideal, for with the absence of convection and mechanical currents the reactants reach each other solely by diffusion and in the ordinary way the precipitate remains in the spot where it is produced. Were it not for the fact that in different kinds of gels the same precipitation reaction produces different effects it would appear to be safe to regard such reactions in gels as an indication of the course normally taken by ionic reactions of double decomposition—a problem which is difficult to study kinetically in any other way—but such a conclusion must be regarded as doubtful at present. We cannot be sure whether the gel modifies the ordinary course of the reaction or whether it reveals the ordinary course of the reaction, which is normally obscured by the turbulence of the mixed solutions. The fact that reactions taking place in capillary spaces in the absence of a gel generally exhibit the same peculiar effects as are observed when the reaction occurs in a gel lends colour to the latter supposition. In that case, the study of chemical reactions in gels becomes a matter of fundamental importance, but it is in any case a subject of very great interest, because chemical reactions in biological structures, whether living or dead, will normally be of this type.

The usual way to study these reactions is to prepare a sol of, say, 10 per cent. gelatin or other suitable colloid, mix it with an equal volume of a dilute solution of one of the reactants, pour the mixture into a series of test-tubes so that they are about one-half or two-thirds full, allow it to set to a gel and then pour more concentrated solutions of the other reactant into the upper part of each test-tube. Alternatively the gelatin-electrolyte mixture may be poured on to a plate of glass and, after it has set, drops of the other reactant may be placed on the gel. Diffusion and reaction then spread radially instead of in one direction. The former method is generally more serviceable.

In either case it takes several hours or even days for a length of a few centimetres of precipitate to form, but the important feature is that the precipitate has a structure. In general, four types of structure may be produced, viz. continuous, discrete, cellular, and periodic. The latter two, in particular, are of the greatest interest in connexion with biological structures.

Precipitates of *continuous structure* consists of small particles uniformly distributed so that the whole mass appears homogeneous to the eye. They are generally produced when the precipitate is a particularly insoluble substance, such as silver chloride or barium sulphate. Sometimes, when examined under sufficiently high magnification, apparently continuous precipitates may be seen to be in reality cellular or periodic. *Discrete structures* consist of particles of relatively large size separated by considerable spaces and may contain well-formed crystals having a length of a few millimetres. They are more common in gels of silicic acid than in gelatin gels, probably owing to the absence of protective effect. *Cellular structures* consist of a network of honeycomb-like cells of precipitate enclosing the clear gel. They are frequently obtained with a large number of precipitates and in many gels, generally when the reacting solutions are dilute. *Periodic structures* contain definite bands or rings of precipitate separated by clear spaces which often contain the precipitate as a discrete structure.

These periodic structures have aroused very widespread interest and it is proposed to discuss them in some detail in the subsequent pages of this chapter.

Periodic Structures

The Liesegang Phenomenon. In 1896, Liesegang (*Phot. Archiv.*, 1896, 221) observed the following peculiar phenomenon. If a drop of fairly concentrated silver nitrate solution be placed on a sheet of gelatin impregnated with potassium dichromate, reaction between the electrolytes takes place in the gelatinous medium with precipitation of silver dichromate. Under these conditions, however, the precipitate is not continuous, but forms a series of concentric rings separated by clear spaces in the gel.

Such periodic formations are customarily called " Liesegang rings," and Liesegang himself was quick to realize the importance of the phenomenon in the explanation of laminated biological and geological structures and suggested that agate-like formations were caused by the diffusion of metallic salts into a mass which was formerly gelatinous silicic acid. The periodic formation is by no means restricted to silver dichromate, nor need gelatin be the reaction medium.

Most subsequent experimenters have employed the method of allowing the gel, containing a small quantity of one of the reacting salts, to set in the lower part of a test-tube and then pouring a more concentrated solution of the other reactant on top. In this case the periodic structure takes the form of a series of bands or discs of precipitate corresponding with a thin strip cut radially from the ring formation. Fig. 19 is an example of this type of periodic structure formed by allowing silver nitrate to diffuse into gelatin containing potassium dichromate.

A considerable amount of experimental work has been per-

FIG 19 —Liesegang Rings. Silver chromate in gelatin.

formed to determine the effect of various conditions on the process. In general, the ring formation in any one system is dependent on the concentration of both reactants and of the gel. Within limits, the distance between the bands decreases with increasing concentration of reactants and gel and the best rings are usually produced by employing a concentrated diffusing agent and a dilute reactant in the gel. In almost every periodic structure of this type the distance between successive bands increases in conformity with a geometrical series. For quantitative investigations of this kind the rhythmic banding of magnesium hydroxide, produced by diffusing ammonia solution into gelatin or agar gels containing magnesium chloride, described by Popp (*Kolloid-Z.*, 1925, **36**, 208) is exceptionally suitable. This is well shown by Fig. 20, which is a photograph of this system taken by the author. The bands are exceedingly sharp and are separated by wide spaces of clear gel.

The magnesium hydroxide system lends itself admirably to quantitative analysis of the bands and the clear spaces and an analytical investigation on these lines was undertaken by Hedges and Henley (*J. Chem. Soc.*, 1928, 2720). The structures were prepared by diffusing concentrated ammonia solution into 7·5 per cent. gelatin gel containing 5 per cent. of crystallized magnesium chloride, thus ensuring sufficient rigidity of the gel for cutting and handling. In order to remove the gels from the test-tubes, a groove was cut so as to encircle completely the tube near the lower, closed end, and the end of the tube could then be removed as a cap by applying a red-hot splinter of glass to the groove. The tube was then dipped for a second into water at about 80°, which had the effect of loosening the gelatin adhering to the glass, and the entire structure could readily be blown out of the tube. All other methods of removing the gels failed because of the strong adherence of gelatin to glass. Agar gels can conveniently be removed by cracking open the tubes. The structures were cut up into bands and clear spaces by means of a safety-razor blade. The white bands were found to be fairly hard and relatively heavy and were readily detached in the form of pastilles from the intervening gel spaces. The analyses showed that the amount of precipitate contained in the bands is twelve times as great as that present in an equal weight of the clear space, whereas the chloride content of the bands is less than one-third of that of the clear space. The equivalent

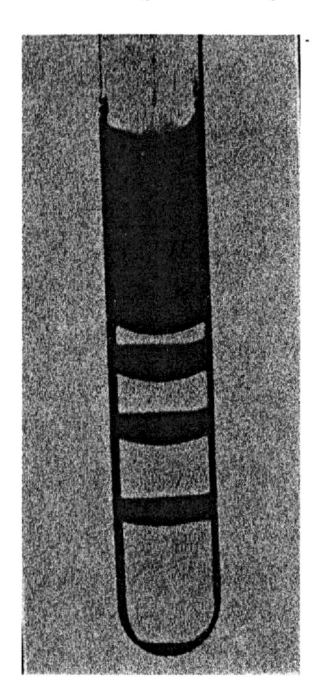

FIG. 20.—Bands of magnesium hydroxide in agar, viewed in transmitted light.

ratio $\tfrac{1}{2}$Mg : Cl is 14·0 : 1 in the bands and 1 : 2·8 in the clear space.

Some interest has been attached to the effect of light on the formation of these structures, but a review of the literature shows that the experimental results are, on the whole, contradictory. As a rule, light does not affect the form of the structure unless light-sensitive materials, such as silver salts, are present. Hatschek prepared stratifications of lead chromate and dichromate in agar and observed, in many cases, wide spaces between sets of bands. These were not developed in the dark, where the normal layers still appeared, and were observed to form diurnally. The ordinary and anomalous bands could be made to form side by side in the same gel by screening one side and illuminating the other. Liesegang has suggested that these effects ascribed to light may be due to temperature differences caused by the irradiation. On the whole, it appears that light may exert a modifying influence in certain cases. Perhaps the main interest of this point is that living organisms are subject to periodic light changes through alternating day and night and in some instances periodic structures which have been related to Liesegang rings may in reality be due to such external periodic changes.

The properties of the gel also play a rôle in the formation of periodic structures. Under ordinary conditions silver chromate readily forms bands in gelatin but not in agar, whilst lead chromate gives bands in agar and not in gelatin, and neither does in silicic acid ; but, by suitably modifying the conditions periodic structures may be obtained in all these instances. At least in some cases the course of events is determined by the other substances present in the gel, and the apparent specificity of the gel is in many cases due to neglect of the alteration of conditions, such as concentration, and on this account many conflicting statements are to be found in the literature. Although the gel has undoubtedly a strong influence in modifying the results it does not seem feasible to entertain hopes of tracking the periodicity to the gel itself. In fact, it may be stated immediately that the presence of a gel is not even necessary. Rings of silver dichromate can be formed by placing a moistened silver nitrate crystal on a glass plate on which a thin film of potassium dichromate has been allowed to dry. Bands of lead iodide have been produced by allowing solutions of lead nitrate and potassium iodide to diffuse into each other in the capillary space between a microscope slide and cover glass. Dreaper produced bands of many crystalline substances by allowing the reacting solutions to diffuse into each other through a capillary tube. Others have obtained similar results in reaction media of kaolin, sand, pulverized calcium carbonate, filter paper, etc. It should be borne in mind, however, that conditions have to be such as may very well exist inside a gel; it appears that such results are realized when surface forces are given full play, by conducting experiments in thin films of solution or in capillary spaces.

Nevertheless, in practice the gel does play a part in determining the structure. Liesegang found that the gelatin should contain small quantities of acid and gelatose in order to produce well-developed rings of silver chromate. Normally, commercial gelatin contains sufficient gelatose through hydrolysis, and rings do not form unless this is present. Also with agar gels, the rhythmic precipitation of various substances varies with the brand of agar, some specimens giving no bands at all.

An interesting effect is the production of anomalous structures. Sometimes spirals are produced instead of rings, and in other cases the bands may form in twins or in triplets. Cellular structures may be other examples of these anomalies. Observations by Hedges and Henley indicate that spiral precipitation of magnesium hydroxide in agar is not caused directly by helicoidal diffusion, but that layers of precipitate reaching half-way across the test-tube are formed alternately on opposite sides and at different levels; later these join to form a spiral. When the bands become more than about 1 centimetre apart they no longer join up. Such effects might be caused by temperature difference due to the exposure of one side of the tube to a draught.

An extreme case of this anomaly was observed in diffusing magnesium chloride solution into agar containing ammonia, giving a " disc and ring " structure similar to that represented in Fig. 21.

Theories of Periodic Structures. Although more than thirty years have elapsed since the discovery of Liesegang rings, interest in these structures has not diminished, and even at the present time experimental work is proceeding in several parts of the world, and new theories are being continually advanced.

A critical survey of the whole subject up to 1926 has been given by Hedges and Myers in their book on *The Problem of Physicochemical Periodicity*, where a practically complete bibliography is given. Unfortunately, many of the theories can be entertained only in respect of the particular case studied, whilst the large amount of experimental data accumulated makes it quite clear that periodic precipitation is a very general phenomenon. It is probably not an exaggeration to say that, given the right conditions, periodic structures can be obtained from any pair of salts which interact to give a precipitate. It is evident, therefore, that a theory which aims at providing an explanation of such a general phenomenon must be of general applicability, and the possibilities of such a general theory have recently been discussed by Hedges (*Rev. Gén. Colloid.*, 1930, 8, 193).

The ideally comprehensive theory might be applicable not only to periodic structures produced by precipitation, but also to similar structures produced by quite other means such as, for example, the rhythmic figures formed by the crystallization of thin films of pure substances described by Hedges and Myers (*J. Chem. Soc.*, 1925, 127, 2432) and by Hedges (*Nature*, 1929, 123, 837), and the "pectographs" recently investigated by Bary (*Rev. Gén. Colloid.*, 1928, 6, 209). Whether so comprehensive a theory will be evolved remains to be seen in the future, but at present among the numerous theories of the formation of periodic structures by precipitation there are a few which, in virtue of their general nature, have received universal interest.

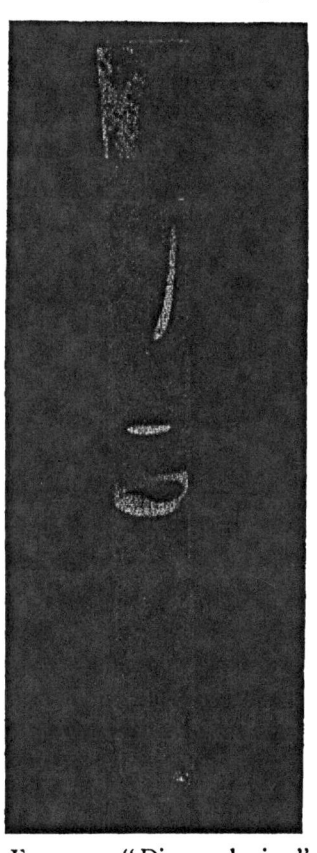

FIG 21.—"Disc and ring" structure of magnesium hydroxide in agar.

Taking as an example the original Liesegang rings formed

when a concentrated solution of silver nitrate diffuses into a gelatin gel containing potassium dichromate, according to the supersaturation theory of Wilhelm Ostwald (*Lehrbuch der allgemeinen Chemie.*, Leipzig, 2nd Edn., 1898), the silver nitrate and potassium dichromate diffuse towards each other, forming a supersaturated solution of silver dichromate, the precipitation of which is delayed. By the time the precipitate has formed at the boundary of the reacting substances, the gel in the immediate vicinity of the precipitate has become impoverished of potassium dichromate, so the advancing silver nitrate has to travel some distance before it can again make the supersolubility product with the dichromate ions. This accounts for the spaces between the bands. This simple theory, the first to be proposed, still enjoys very considerable favour. The chief objections are that bands of lead iodide have been obtained by diffusing lead nitrate solution into an agar gel containing potassium iodide and sown with crystalline lead iodide, where supersaturation should be impossible, and that in a tube already containing a series of bands of precipitate it is possible, by a second diffusion, to obtain a further series of bands independent of the first series. Liesegang and others have pointed out that these objections are not insuperable, owing to different crystallization conditions existing in the gel.

Bradford's theory (*Biochem. J.*, 1916, 10, 169; 1917, 11, 14; 1920, 14, 29, 474) postulates that the spaces between the bands are to be attributed to the adsorbent effect of the colloidal precipitate on the electrolyte in the immediate vicinity. In favour of this view it may be pointed out that precipitates possessing a large specific surface are most addicted to band formation, and periodic structures may often be obtained or not at will by varying the specific surface and consequently the adsorptive capacity. It has not been possible to show, however, that adsorption of the electrolyte does occur to the extent required, and, indeed, the few experiments which have been conducted seem to indicate that the degree of adsorption is far too small to account for the effect.

Hatschek has made the important observation that in many cases of periodic precipitation the actual amount of reaction product in equal volumes of ring and clear space is the same, the chief difference being the size of the particles. Thus the rings contain a large number of small particles and the spaces contain a small number of large particles. This conclusion, however, is certainly not general, as is shown in fact by Hedges and Henley's analyses of the magnesium hydroxide periodic structure. Dhar and Chatterji distinguish two kinds of ring formation; in one class a layer of precipitate is followed by a clear zone, and in the other class a coagulated sol is followed by a zone of peptized sol. They believe that periodic structures are formed only when the gel has a medium peptizing influence on the precipitate.

The fundamental feature of the "diffusion wave" theory of Wolfgang Ostwald (*Kolloid-Z.*, 1925, **36**, 380) is the rôle played by the *soluble reaction product*—potassium nitrate in the reaction between silver nitrate and potassium chromate—which is said by its accumulation to reverse the reaction. It is postulated that in all reacting systems which give periodic precipitates there are at least three principal diffusion waves and that the periodic structure is produced by the interference of these waves of diffusion. The three diffusion waves are due respectively to the outer and inner electrolytes and to the soluble product of reaction. Periodic structures should therefore be most readily obtained in reactions which do not proceed to completion. This is correct, in general, and is witnessed by the difficulty in producing periodic structures of silver chloride or barium sulphate and also by the exceptionally good macroscopic effect obtained with magnesium hydroxide, which is one of the best cases of incomplete precipitation. However, if it can be shown that periodic structures can be obtained in reactions where there is no third diffusion wave or no soluble reaction product, it cannot be maintained that the theory is applicable to those cases. Several objections on these lines have been collected in a criticism of this theory by Hedges (*Kolloid-Z.*, 1930, **52**, 219).

It is often objected that these principal theories do not take into consideration the specific influence of the gel in which the reaction takes place. This is not quite true. On Wi. Ostwald's theory, the gel can affect the velocity of formation of nuclei from the supposed supersaturated solution; on Bradford's theory, the gel may determine the specific surface of the precipitated substance and consequently its capacity for adsorption; and the theory of Dhar and Chatterji is directly concerned with the peptizing influence of the gel. Wo. Ostwald's theory does not consider the specific influence of the gel except in so far as the gel modifies the relative rates of diffusion of the three electrolytes concerned; and dilute gels have little influence on the rate of diffusion of salts. In any case, this is probably not a point of fundamental importance.

There is a distinct advantage in carrying out these investigations in glass capillary tubes in the absence of a gel, rather than by the more usual way of causing the substances to react in a gelatinous medium in a test-tube. The gel can rarely be regarded as an inert medium; it may, and most probably does, undergo unknown reactions with both reactants and products.

The Distinction between Periodic Structures and Periodic Reactions. The formation of Liesegang rings and other periodic structures are often described as "periodic reactions." There has been no reason to suppose that in the formation of these rings the chemical reaction occurs periodically, but only that the final structure shows a spatial periodicity. The author therefore suggests that these formations exhibiting structural periodicity be called "periodic structures," reserving the term "periodic reactions" to denote chemical reactions where the *time-course* of the reaction is definitely periodic, such as those investigated by Hedges and Myers (*J. Chem. Soc.*, 1924, **125**, 604, 1282; 1925, **127**, 445).

This aspect of the matter was investigated by Hedges and Henley by determining whether the Liesegang rings were formed directly in the chemical reaction or whether the periodicity is a secondary phenomenon occurring after formation of the substance. They succeeded in separating the formation of this type of periodic structure into two stages. In the first stage chemically equivalent dilute solutions of the two reagents in sols of gelatin or agar were mixed, and the mixture was allowed to cool and form a gel without production of any precipitate, and in the second stage a concentrated solution of one of the reagents was allowed to diffuse into the gel, whereupon the insoluble substance was precipitated in the form of a periodic structure. It may be concluded from these experiments that the formation of the periodic structure is not the result of a periodic chemical reaction, but is a secondary phenomenon taking place after the completion of the reaction.

As an example, reference may be made to Fig. 22. This structure was made by mixing equal volumes of 1 per cent. agar solutions containing 0·6 per cent. of potassium iodide and 0·6 per cent. of lead nitrate (equivalent weights), respectively. The hot solutions were cooled and allowed to set in a test-tube. On cooling, the lead iodide did not separate out, although its solubility in water was exceeded many times. When 20 per cent. potassium iodide solution was allowed to diffuse into the tube, a ring system was formed which is comparable with that produced by chemical reaction.

In order to eliminate any complications due to the simultaneous production of potassium nitrate in making up the gel mixture, the experiments were repeated in agar gels containing lead iodide in the absence of other electrolytes. These were prepared by dissolving recrystallized lead iodide in 1 per cent. agar solution at 100°, and allowing it to cool. A few trials showed that the cooled gel would hold as much as 0·2 per cent. of lead iodide, although the ordinary solubility in water is only 0·04 per cent. Equally good bands of lead iodide were produced when 20 per cent. potassium iodide solution was allowed to diffuse into these gels.

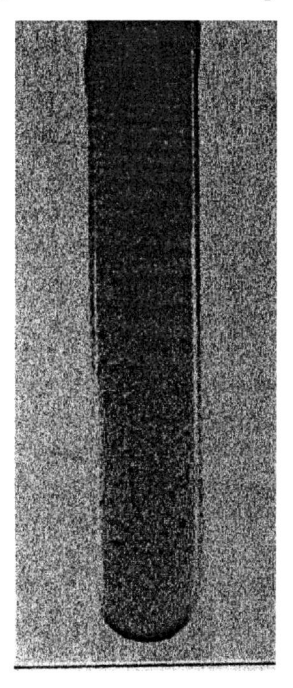

Fig. 22.—Bands of lead iodide in agar formed without chemical reaction.

In this case, chemical reaction has been eliminated entirely, and the only diffusion waves are those due to potassium iodide and lead iodide. There is no soluble reaction product and no third diffusion wave, so it seems that Wo. Ostwald's theory is inapplicable. Since there is the possibility of the formation of soluble complexes such as $KPbI_3$ and K_2PbI_4 when concentrated solutions of potassium iodide are used, the author has since repeated these experiments, using for the diffusing solutions 10 per cent., 5 per cent., 4 per cent., 2 per cent.,

1 per cent., and 0·5 per cent. potassium iodide respectively. In all these cases rings were formed and yet the complexes cannot exist in the more dilute solutions.

Periodic Structures formed by Coagulation. When the experiments which have just been described were carried out it was believed that the insoluble reaction product is first formed as a colloidal sol, which remains protected by the gel of agar or gelatin, and that the rings are formed by a periodic coagulation process effected by the excess of diffusing agent. It may therefore be expected that periodic structures can be obtained by simple diffusion of a coagulating electrolyte into a sol contained in a gelatin or agar gel. To test this view, experiments were conducted on the diffusion of coagulating electrolytes such as ferric chloride and aluminium sulphate into sols of arsenious sulphide and ferric hydroxide contained in agar gels. In these experiments periodic coagulation structures were obtained.

More recently, Hedges (*J. Chem. Soc.*, 1929, 2781) has obtained periodic coagulation structures by a simpler method. A 1 per cent. sol of arsenious sulphide was contained in a series of capillary tubes, open at one end, which were then immersed in solutions of ferric chloride varying in concentration from 1 to 30 per cent. No gel was present. In these capillary experiments definite bands of precipitate were not formed, but the coagulum consisted of an undulatory filament, the distance between successive peaks of the waves being generally 2–5 mm. Fig. 23 is a photograph of one of these structures taken at the author's suggestion in the laboratories of the Wool Industries Research Association by Miss M. H. Norris, under the direction of Dr. S. G. Barker. This is a new kind of periodic structure and may be of interest in connexion with the waviness of many natural colloidal fibres such as wool and hair. It appears, therefore, that periodic structures can be produced by simple coagulation.

In the ordinary Liesegang phenomenon it is by no means certain that the precipitate is first produced in colloidal form. The question whether the insoluble substance which is finally precipitated in the form of periodic structures such as Liesegang rings exists primarily as a highly supersaturated solution or as a colloidal solution which is protected by the gelatinous reaction medium has already received some attention. The experiments of Sen and Dhar (*Kolloid-Z.*, 1924, **34**, 270) and of Chatterji and Dhar (*Trans. Faraday Soc.*, 1926, **23**, 233 ; *J. Indian Chem. Soc.*, 1928, **5**, 175) suggest that when a dilute solution of potassium chromate in gelatin at the concentrations ordinarily used in making Liesegang rings is treated with an equivalent of silver nitrate, the silver chromate, which is not precipitated under these conditions, remains in colloidal solution and shows little diffusion or electrical conductivity. On the other hand, the diffusion and electro-conductivity experiments of Williams and Mackenzie (*J. Chem. Soc.*, 1920, **117**, 844), of Bolam and Mackenzie (*Trans. Faraday Soc.*, 1926, **22**, 151, 162), and of Bolam (*ibid.*, 1928, **24**, 463) favour the view that the silver chromate exists as a highly supersaturated molecular solution.

It is perhaps unwise to attempt to distinguish sharply between the condition of a little-understood, highly supersaturated, "molecular" solution and a highly disperse, colloidal solution; but if the solutions obtained by mixing the equivalent dilute reactants in the gel can be regarded as molecular, the second stage occurring on diffusion of the excess of concentrated reactant becomes a salting-out process by a common ion rather than one of coagulation. Hedges and Henley also showed that 1 per cent. agar gel would hold as much as five times the amount of lead iodide corresponding to the ordinary solubility in water at the temperature ; it was assumed that the excess was present as a sol protected by the gel of agar, but even on that assumption it must still be supposed that at least 20 per cent. of the lead iodide must be molecularly dissolved. It appears, therefore, that the bands of precipitated lead iodide produced when 20 per cent. potassium iodide solution is allowed to diffuse into such a mixture may be formed by a periodic salting-out of lead iodide by potassium iodide rather than by a coagulation process.

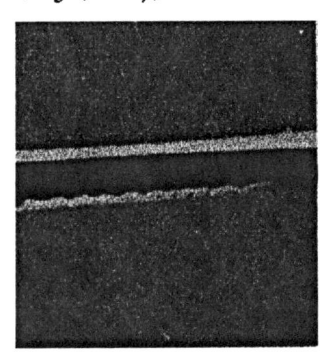

FIG 23.—Undulatory filament of coagulated arsenious sulphide.

Periodic Structures formed by Salting-out. Experiments

have been conducted by Hedges (*J. Chem. Soc.*, 1929, 2779) to determine whether such periodic structures can be formed by a pure salting-out process in the absence of colloids, and thus to differentiate the phenomenon from coagulation as understood by colloid chemists. For this purpose, hydrochloric acid was allowed to diffuse into sodium chloride solutions, for example, and the experiments were conducted in capillary tubes in the absence of a gel.

The procedure adopted was to allow the salt solution to rise in a capillary tube of about 0·5 mm. diameter, seal off the top

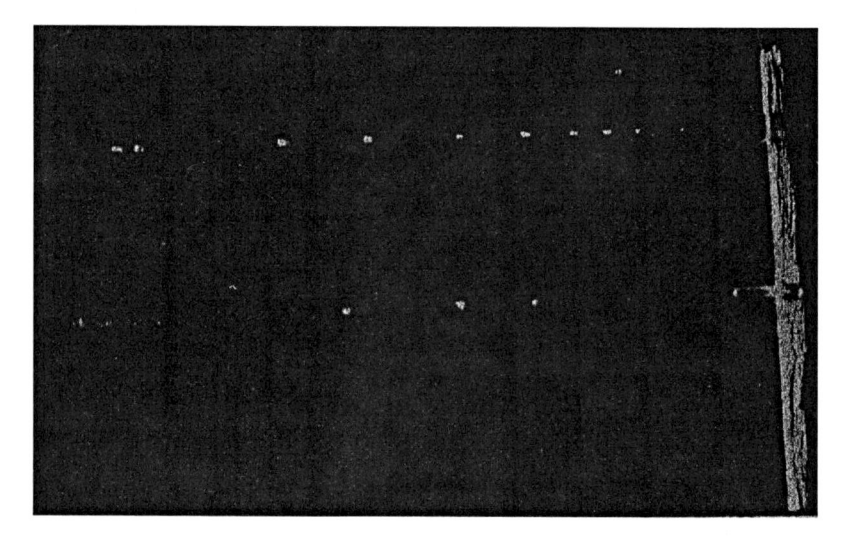

FIG 24.—Periodic structures of sodium and potassium chlorides formed by salting-out.

end of the tube, leaving the lower end open, and place it in a horizontal position in a corked test-tube containing the acid, which could thus diffuse into the capillary.

An example of the results is afforded by the upper tube in Fig. 24, where the capillary tube contained 30 per cent. sodium chloride solution and the outer diffusing liquid was concentrated hydrochloric acid. After some days, eight bands of sodium chloride had appeared in the tube. The distance between successive bands, starting at the open end (right in Fig. 24) of the tube, were 4·0, 6·0, 8·5, 11·5, 10·5, and 18·0 mm. (The last band became displaced before the photograph was taken.)

It will be seen that the spacing increases throughout the diffusion as in the ordinary Liesegang phenomenon; similarly the thickness of the successive bands increases from 1·0 to 1·5 mm. Similar periodic structures were obtained in the salting-out of concentrated potassium chloride solutions by hydrochloric acid (lower tube in Fig. 24) and of barium nitrate by nitric acid.

By eliminating both chemical reaction and colloid-chemical processes, these experiments are perhaps the simplest examples of the formation of periodic structures by diffusion yet investigated, and they appear to be at variance with some of the best-known theories of the formation of Liesegang rings. Bradford's adsorption theory supposes that the spaces between the rings are caused by the adsorption of the electrolyte contained in the gel by the adjacent layer of precipitate; such adsorption will more probably occur if the precipitate is in a highly disperse state and bands can in some cases be obtained or not, at will, by varying the conditions affecting the dispersity of the precipitate. In the salting-out experiments described above, the " bands " of precipitate consisted sometimes of large single crystals, but generally of aggregates of a few relatively large crystals. The spaces between were perfectly devoid of solid. Adsorption would be negligible on crystals of such extremely small specific surface. The diffusion-wave theory of Wo. Ostwald ascribes the most important rôle to the diffusion of the soluble reaction product. In the experiments which have just been described, however, this third diffusible product has been eliminated; there are only two diffusion waves to be considered—those of hydrochloric acid and of sodium chloride. Another case in which the diffusion wave theory appears to be inapplicable is in the formation of bands of ammonium chloride by the interaction of dry hydrogen chloride and ammonia gas (Hedges, *J. Chem. Soc.*, 1929, 1848). In the reaction

$$NH_3 + HCl = NH_4Cl$$

there is only one reaction product.

The considerations outlined in the preceding pages tend to annihilate a sharp distinction between the coagulation of a sol and precipitation by exceeding the supersolubility limit, and bring Wi. Ostwald's supersaturation theory into line with the coagulation theory of periodic structures. The essential condition for periodicity appears to be the existence of some *critical condition* determining a change which proceeds to completeness once the critical value is reached. In an investigation of the cause of periodic phenomena in electrolysis and in the chemical passivity of iron, it has been shown (Hedges, *J. Chem. Soc.*, 1929, 1028) that these phenomena are determined by the rate at which a certain critical concentration of ions can be reached by diffusion. In the formation of periodic structures by diffusion and precipitation it seems probable that a critical concentration is again involved; at least, the coagulation of sols by electrolytes is associated with a definite threshold concentration. A theory of the formation of periodic structures based on the idea of a critical concentration value for coagulation of the primarily formed sol has been advanced by Freundlich (*Colloid and Capillary Chemistry*, 1926, p. 735) and appears to be capable of extension to include structures formed by salting-out.

Natural Periodic Structures. Natural periodic structures which resemble the Liesegang phenomenon in appearance are frequently encountered in both geology and biology, but whilst there is much evidence to show that banded mineralogical specimens such as agates can be traced to the diffusion of salts, and can thus be correlated with the formation of artificial periodic structures, it is not always safe to apply these principles to the biological specimens. This is partly because there are so many external periodic conditions which can account for their formation, such as alternating day and night, winter and summer, sleep and rest, intermittent feeding, etc. Typical examples are the form of grains of starch and such layered concretions as gall-stones and urinary calculi. Attempts have been made to apply the Liesegang principle to the markings on butterflies' wings and on the shells of certain molluscs, and the rings in beetroot slices and in sections of trees have also been considered from this point of view. Surely the rings in tree-trunks are definitely caused by an external seasonal periodicity, and enthusiasm has been carried beyond the bounds of prudence when it has been suggested that the stripes on tigers and zebras may be a glorified Liesegang phenomenon.

CHAPTER XVII
COLLOIDS AND CHEMICAL REACTIVITY

Direct Reactions of Colloids. The literature reveals a certain amount of dissension of opinion regarding the velocity of reaction of colloids with chemical reagents, some authors regarding colloids as sluggish, whilst others state that they react rapidly. On the whole, however, the chemical reactivity of colloids is consistent with their state of subdivision ; that is to say, they react quickly when compared with the substance in massive form, but sluggishly when compared with a true molecularly disperse solution or with a solution of ions.

Particularly rapid reactions are the almost instantaneous decolorization of arsenious sulphide hydrosol by alkali and the almost immediate precipitation of ferric hydroxide sols as sulphide by hydrogen sulphide. On the other hand, a sol of ferric hydroxide is first precipitated on the addition of hydrochloric acid and dissolution occurs subsequently ; obviously, the velocity of flocculation is greater than that of dissolution. Another instance of sluggish reaction is afforded by silver sols, which dissolve only slowly in nitric acid, where the massive metal would be rapidly dissolved. In cases such as this, however, it is very probable that the particles are surrounded by some envelope that opposes dissolution. For example, an envelope of gas is particularly likely in silver sols which have been prepared by Bredig's electrical disintegration method and, as already noted in other cases, the surface of the particles is rarely regarded as having the same composition as the inside. Rapid dissolution ensues when once the protective layer has been broken down. The properties of these and other protective layers on metals have been summarized by Hedges (*Chem. and Ind.*, 1931, **50**, 21).

Even in some cases of dissolution without chemical reaction in the ordinary way the same sluggish behaviour, probably due to some protective film, is encountered. Hydrosols of sulphur, prepared by leading sulphur vapour into water, are not acted on by carbon disulphide, although the sulphur which slowly precipitates on keeping the sols dissolves in carbon disulphide readily.

Catalysis. More interest is attached to the effect of colloids on other reactions. Heterogeneous catalysis involves adsorption in some form or other and, other things being equal, specific surface is a predominating influence. The activity of colloidal catalysts is therefore generally very high and metal catalysts when in the colloidal state become extraordinarily active, mainly, although not entirely, because of the great increase in specific surface.

Enzymes are typical examples of colloidal catalysts. Most animal matter and a good deal of vegetable matter contains an enzyme known as *catalase*, which decomposes hydrogen peroxide. The catalytic decomposition of hydrogen peroxide by colloidal noble metals has been studied in great detail by Bredig, who found the decomposition very similar to the action of enzymes. He refers to colloidal metals as " inorganic ferments." In these reactions the age of the sol is of importance. Hedges and Myers (*J. Chem. Soc.*, 1924, **125**, 1282) found that under certain conditions the rate of decomposition of hydrogen peroxide both by colloidal metals and by catalase becomes periodic.

Catalysis does not depend entirely on specific surface, but also on the existence of " active centres " in the surface, the whole of the catalyst not being active. That the nature of the surface as well as its extent plays a part in catalysis by colloids is demonstrated by some work by Galecki and Krzeczkovska (*Bull. Acad. Polonaise*, 1925, A, 111), who showed that gold sols prepared above 50° were less active in decomposing hydrogen peroxide than those prepared at lower temperatures. From parallel ultramicroscopical observations and measurements of viscosity and mobility in an electric field it was concluded that activity does not depend only on the active surface of catalyst, but also on the nature of that surface, as affected by its method of formation.

In some of these so-called catalytic reactions the colloidal catalyst does undergo chemical change. For example, according to Koller-Aeby (*Kolloid-Z.*, 1928, **45**, 371), in the catalytic decomposition of hydrogen peroxide by colloidal silver the catalyst does not remain unchanged, and the change undergone depends on the concentration of the sol. In general, the sol changes from a dark brown to a yellowish-brown, but at certain concentrations the colour vanishes completely and the solution fails to show a Tyndall cone. It is suggested that in the reaction silver oxide or other oxidation products are formed.

Other catalytic reactions under the influence of colloidal platinum which have received special attention are the union of oxygen-hydrogen gases and the oxidation of aqueous chromous chloride to chromic chloride with evolution of hydrogen.

The colloidal state does not without exception confer enhanced catalytic activity, however. A recent investigation by Gibbs and Liander (*Trans. Faraday Soc.*, 1930, **26**, 656) has shown that nickel aerosols prepared by vaporization of nickel in a continuous direct-current arc or by heating the vapour of nickel carbonyl, and nickel aerogels prepared by the condensation of these aerosols, have little or no catalytic effect on the reduction of carbon monoxide or ethylene by hydrogen.

Hydrophilic Colloids. The constitution of hydrophilic colloids is almost always so complex that very little can be said about the reactions they undergo. The effect of such substances on other reactions has received a little study, however, and as a rule the presence of a hydrophilic colloid slows down any change involving a change of state. Thus, gelatin has very little influence on the rate of hydrolysis of ethyl acetate, but retards greatly the rate of dissolution of metals in acids; in general, truly homogeneous reactions are not appreciably affected, but heterogeneous reactions involving surfaces of separation are retarded. This is conceivably due to the fact that the colloid is adsorbed at such interfaces and interferes with the processes of diffusion and reaction; the changes are too great to be explained by the increase in viscosity of the system.

Gelatin not only retards chemical reactions involving the appearance of a new phase, such as the catalytic decomposition of hydrogen peroxide, but also retards changes of state which need not involve a chemical reaction, such as the change of plastic sulphur to the crystalline allotrope and the change of yellow mercuric iodide to the red form. When two gelatin solutions containing respectively dilute mercuric chloride and potassium iodide are mixed the precipitate is yellow and does not become pink until after some hours. If the reaction is carried out in the absence of gelatin the yellow modification can be seen only momentarily, being transformed to the red variety in a fraction of a second.

Friend and Dennet (*J. Chem. Soc.*, 1922, **121**, 41) found that the velocity of dissolution of iron in dilute sulphuric acid is lowered considerably in presence of gelatin, and later work has shown that this effect is quite general in the corrosion and dissolution of metals. Some experiments carried out by Hedges and Myers (*J. Chem. Soc.*, 1924, **125**, 620) illustrate the magnitude of the changes brought about by quite small quantities of the hydrophilic colloid. In the case of aluminium and N- sodium hydroxide, 0·02 per cent. of gelatin lowered the velocity of dissolution by 10 per cent. With zinc and N- hydrochloric acid, the velocity was reduced 50 per cent. by 0·02 per cent. of gelatin. The small quantities involved are characteristic of adsorption processes.

Periodic Reactions. An interesting effect of colloids on the dissolution of metals is that under certain conditions the rate of dissolution becomes periodic, increasing and decreasing at regular intervals of time, which may be less than one second or more than one hour. Hedges and Myers (*J. Chem. Soc.*, 1924, **125**, 604) have observed the periodic dissolution of zinc, iron, aluminium, magnesium, manganese, and sodium in a number of reagents, including acids, alkalis, neutral salts, water, and alcohol, showing that, given the right conditions, the phenomenon is quite a general one. In Fig. 25 are reproduced some automatic records of the variation of pressure of the evolved hydrogen with time, brought about by the addition of colloids to aluminium amalgam reacting with water.

The special conditions required have never been completely elucidated, but it is quite certain that the periods are obtained only in the presence of some foreign substance, which may be present in extremely small amount ; in some cases it was evident that the glass reaction vessel was capable of furnishing the requisite trace of " catalyst," whilst in others the addition of colloids in so small an amount as 1 part in 1,000,000 brought about periodicity. It was also found that the periodicity was stopped by the addition of 1 part in 1,000,000 of chloroplatinic acid. The minuteness of the quantities involved rendered the work very difficult, but is in itself characteristic of biological systems. A particularly interesting observation was the effect of the addition of certain poisons. Small quantities of formaldehyde, potassium cyanide, or sodium arsenite decreased the frequency of the periods very greatly, and larger amounts stopped the periods completely. Presumably their influence lies in their effect on the colloid.

An interesting feature not only of periodic chemical reactions of this type, but also of the periodic structures discussed in Chapter XVI, is their association with colloids, for periodicity of both types is very characteristic of biological systems, which consist practically entirely of colloidal matter. The study of these reactions and structures may therefore lead to a better understanding of reactions in the living organism, and the relations

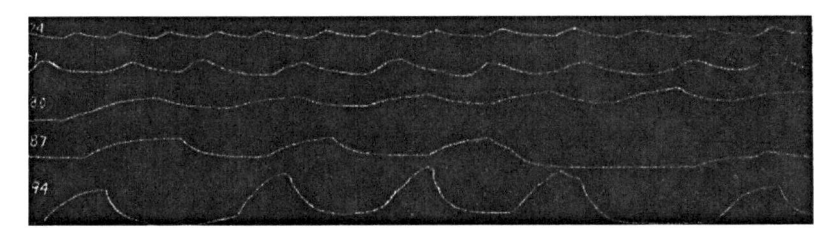

FIG. 25 — Automatic records of the rate of evolution of hydrogen from aluminium amalgam and water in the presence of various colloids.

between physico-chemical and biological periodic phenomena have recently been discussed by Hedges (*J. Soc. Chem. Ind.*, 1930, **49**, 121T).

Even when colloids have not definitely been added to the reacting system a product may first appear in the colloidal state. Periodicity appears to be generally, if not always, associated with the formation of a film, and such films are often produced under conditions which would favour their appearance as a colloidal gel.

Experiments on the periodic dissolution of a copper anode in various chloride solutions have shown that the frequency depends entirely on the rate of accumulation of chlorine ions and appear to leave no escape from the conclusion that dissolution of the film does not occur until a certain critical concentration is reached, but that when reaction occurs in each " pulse " it is complete. This is analogous to the " all-or-nothing " law so familiar in physiology, according to which a system does not respond to a stimulus until this reaches a threshold value, but the response is complete when that value is reached. Independent evidence of the existence of such critical concentrations has been obtained.

Physiologists use a similar view to explain the periodicity of breathing; respiration occurs every time the carbon dioxide content (hydrogen-ion concentration) of the blood reaches a threshold value.

Experiments on the periodic electrodeposition of metals through secondary reaction afford strong support for the critical concentration view, and are difficult to explain in any other way. For example, in the electrolysis of mercury-potassium cyanide solutions under certain conditions the potassium liberated at the mercury cathode reacts alternately with water molecules and with the complex $Hg(CN)_4''$ ions or undissociated salt, the final effect being a periodic evolution of hydrogen alternating with deposition of mercury. The supposition here is that reaction of mercury with the complex ions does not take place until the ions reach a certain critical concentration by diffusion. This mechanism is similar in outline to the older idea that the heart-beat is caused by alternate reaction with or adsorption of calcium and potassium ions, although that view does not appear to be greatly favoured by present-day physiologists. It may also be pointed out that critical concentrations of ions are required for the coagulation of colloid systems and that these systems are the most suitable for the observation of periodic phenomena, both in time and in space.

CHAPTER XVIII

SOME APPLICATIONS OF COLLOID PRINCIPLES

The applications of the principles, which have been discussed in the preceding chapters, to laboratory practice and to technical processes are so numerous that a separate volume might very well be devoted to their consideration. In the present chapter a few examples have been chosen and are discussed briefly in order to give some idea of the practical importance of the study of colloids.

Smokes and Fumes. The problem of smoke abatement in industrial areas has already aroused some interest and it is to be hoped that in the future that interest will become more widespread. The prevention of the smoke nuisance is important from two points of view, for not only does the darkening of the sky caused thereby and the subsequent deposition of soot cause untold harm to human health and enjoyment and to animal and plant life, but in the solid particles ejected from chimneys are many useful substances which are often worth the cost of collecting. The smoke from chemical works often contains particles of definitely poisonous substances, which on any account must not be allowed to pollute the atmosphere.

Various methods are employed in industry for the elimination of these solid and liquid particles, such as filtration, washing by sprays of water, passing the gases through settling chambers where the velocity and temperature of the particles are so reduced that they settle under gravity, arresting by means of baffle plates, and electrical precipitation.

The electrical precipitation of smokes and fumes involves the principle of cataphoresis and was first put on a technical basis by Cottrell. The particles are first brought into contact with fine points having a high electrical potential, so that they become charged with the same sign as that of the fine points. The charged particles are then attracted to and discharged upon a large plate electrode having the opposite sign and a low potential and generally made of lead or iron. The precipitated material is readily shaken or washed off the electrode if solid and if liquid can be collected simply. The high potential point electrodes are generally made of mica or asbestos and may be charged to 100,000 volts.

The process is employed in many silver, zinc, and arsenic plants and also for the recovery of sulphuric acid where this is present in the fumes from a process. The same principle has also been applied to the dehydration of petroleum-water emulsions which are often difficult to separate into their constituent phases.

These processes recall also an application of electro-osmosis to the dehydration of peat, the water suffering migration under the influence of an applied electric field and being continuously removed at one of the electrodes.

Clays and Soils. According to Oakley (*J. Chem. Soc.*, 1927, 2819), the general behaviour of washed clay suggests that it is an insoluble or colloidal weak acid with a dissociation constant of approximately 10^{-8}, in equilibrium with bases in solution, the salts formed being equally insoluble.

Naturally occurring clays are rather complicated mixtures of hydrated silicates of the alkalis, alkaline earths, iron, and aluminium, generally with some sand and organic matter. Colloidal silica, silicates, and metallic hydroxides with colloidal organic matter usually form a sheath round the particles and account for many of the properties of the clay. The amount of inorganic colloidal matter rarely exceeds 1·5 per cent. ; if the content is low the clay is sandy and has a low plasticity and is said to be weak or lean ; if high, the clay will be sticky rather than plastic and is said to be fat. In order to provide the correct degree of plasticity for ceramics, therefore, the proportion of colloidal matter in clay is important.

The colloidal content of the clay is also responsible for the shrinkage undergone during firing, highly colloidal clays shrinking most and often cracking on being dried. Such clays have therefore to be mixed with a suitable proportion of lean clay and often they are preheated before mixing. This treatment has the effect of partly dehydrating the colloidal constituents, and on mixing with water they do not immediately become rehydrated. Indeed, the slow rehydration of clay is familiar to all who have struggled with a garden ; after a prolonged period of drought the soil does not become sticky even after heavy rain, but during a spell of wet weather the soil becomes so sticky as to be almost unworkable.

Adsorption is an important property of clays and is utilized in determining the content of colloidal matter. The study is very complicated, because there are so many possibilities of exchange adsorption in so complex an adsorbent. In the adsorption of acid or neutral salts, sodium, potassium, and magnesium from the clay may be released equivalent in amount to the adsorbed cation, and in the addition of alkaline solutions the free alkali may react with the colloidal silica and the cations displaced from the clay may form soluble salts which may be adsorbed by constituents of the clay. These reactions are of importance in studying the fertilization of the soil and it is said that the addition of lime to a soil not only neutralizes acids but renders the potassium of the clay available for the plant.

A recent investigation of the soil colloids by Puri (*Mem. Dept. Agric. India*, 1930, **11**, 1, 39) reveals interesting data. When a soil is exhaustively treated with $N/20$-hydrochloric acid and washed free from excess, the amount of hydrogen-ion retained by it is a characteristic constant for the particular soil. A part of the hydrogen-ion content can be replaced by any other cation by exhaustive treatment with a neutral salt, and the free acid thus liberated is equivalent to the amount of cation entering the soil complex. The " acidoid " or soil completely saturated with hydrogen ions is a complex acid, exhibiting the usual properties of acids, and having a dissociation constant which is close to that of acetic acid. The main result of the investigation is to show that the acidoid possesses the reactions characteristic of a true tribasic acid. The first hydrogen ion can be replaced by exhaustive treatment with a neutral salt and a neutral substance is thereby produced ; the second hydrogen ion can be replaced by heating the suspension with an alkaline earth carbonate in a current of air ; and the third hydrogen ion is replaced by using a large excess of alkali. It is proposed to call the soil completely saturated with hydrogen ions " clayic acid " and the salts derived therefrom " clayates." The dispersion coefficients or percentage of the clay which can pass into suspension on being left in contact with water were determined for clayic acid and a number of clayates. Lithium and sodium salts gave considerably higher values than any other. All good soils give a comparatively low dispersion coefficient and all barren alkali soils give a high value. Soils rich in replaceable sodium give with a high dispersion coefficient a very low rate of percolation to water, and whilst treatment with gypsum increases the permeability it lowers the dispersion coefficient. Highly acid soils give a very low dispersion coefficient and a very high rate of percolation. Irrigated soils give a comparatively higher dispersion coefficient than unirrigated soils.

Another recent investigation involving the stability of clay suspensions is due to Foerster (*Kolloid-Z.*, 1930, **52**, 160), who finds that the rate of settling of fine suspensions of clay in water is influenced by the presence of electrolytes and particularly by alkalis. Sodium hydroxide increases the stability of the suspension, but calcium hydroxide causes relatively rapid sedimentation. Calcium and magnesium sulphates and chlorides settle the suspensions at a concentration of only $N/1000$. Thus, the suspensions settle more readily in hard than in soft waters. In technical practice, the addition of milk of lime or calcium sulphate causes clay turbidities to clear completely in twelve hours.

In addition to the inorganic colloids which have been described, soils also contain organic colloids, the most important of which is humus or humic acid. This is an acidic substance of very complex chemical nature formed by the decomposition of organic matter ; it is a good culture substance for bacteria and micro-organisms beneficial to the soil, forms a store of nitrogenous matter, acts as a protective colloid for the inorganic colloidal constituents, and adsorbs moisture and plant foods such as nitrates and phosphates. The adsorption of moisture is an important point, for the amount of moisture retained by soil in dry weather is proportional to the content of colloidal matter. Humus is a negatively charged colloid and is therefore coagulated by multivalent cations in the soil, particularly those of calcium, iron, and aluminium.

Colloidal Graphite. Colloidal carbon is known in the form of tars, but there are two preparations of colloidal carbon or graphite which have achieved technical importance as lubricants. These are known respectively as " Waterdag " and " Oildag " and consist of colloidal graphite in water and in oil. Waterdag is made from the extremely finely divided graphite found in the electric furnaces employed for the manufacture of carborundum by masticating it with water and gallotannic acid, when stable sols are produced, containing up to 1 per cent. of graphite. The finely divided graphite does not disperse directly in oil and in the preparation of oildag it is first necessary to make a paste of graphite and tannin with water and then gradually substitute oil for the water. The oil paste is then diluted by further addition of oil to the desired consistency.

Qualitative and Quantitative Analysis. A knowledge of the behaviour of colloids and of their conditions of formation and destruction is very important in analytical practice and also for a proper understanding of the theory of qualitative and quantitative analysis.

The passage of a precipitate through the filter paper on washing is often a troublesome result of peptization, and in other cases devices must be adopted to prevent a substance being formed as a sol instead of a precipitate. The means available for the coagulation of such sols are somewhat limited, however, because it is undesirable to introduce fresh ions into the solution being analysed. Multivalent ions can rarely be used as coagulators and recourse is generally had to ammonium salts. One of the functions of the ammonium chloride introduced in the ordinary course of analysis is to aid the precipitation of the otherwise colloidal sulphides of the zinc group. One of the objects of removing organic matter before testing for metals is because such matter may act as a protective colloid and give rise to sols where precipitates are required.

On the other hand, sol formation can occasionally be put to practical advantage in qualitative analysis. For example, the apparent solubility of zinc hydroxide in excess of sodium hydroxide, when separating zinc from manganese, is almost certainly a case of peptization to a sol, for, when the co-called sodium zincate solution is passed through an ultra-filter the zinc hydroxide is retained. It is a common observation that, whilst warming the sodium hydroxide assists the dissolution of the zinc hydroxide, boiling the solution is to be avoided, as the zinc hydroxide is reprecipitated thereby; this is most probably due to coagulation of the sol. Chromium hydroxide likewise dissolves in sodium hydroxide solution by peptization and is reprecipitated on boiling the sol. The peptized chromic hydroxide has an interesting protective effect on ferric hydroxide, which is not peptized under the same conditions. If, however, sodium hydroxide is added to a mixture of chromic and ferric salts, the chromium being in considerable excess, ferric hydroxide is not precipitated, but remains as a sol. If the ferric salt is in excess, the addition of alkali precipitates ferric hydroxide, which adsorbs chromium hydroxide, carrying it out of solution.

In volumetric analysis the presence of hydrophilic colloids may alter or obscure the end-point as determined by the colour-change of indicators. An investigation of this effect has been made by Gutbier and Brintzinger (*Kolloid-Z.*, 1927, 41, 1), for the reaction between $0 \cdot 1 N$- hydrochloric acid and $0 \cdot 1 N$- sodium hydroxide. In the cases where a difference in end-point was noticed, the amount of deviation increased progressively with increasing amounts of colloid, and the result depended on whether the neutral point was approached from the acid or alkaline side. The colloids studied were gelatin, gum arabic, and dextrin, and the results obtained with the indicators were as follows : azolitmin and neutral red underwent the colour-change at the normal value of p_H ; phenolphthalein and rosolic acid behaved normally in the titration of acid, but the end-point was premature in the titration of alkali ; alizarin behaved normally in the titration of acid, but the end-point was delayed in the titration of alkali ; Congo-red and methyl orange gave a premature end-point in the titration of acid and a delayed end-point in the titration of alkali. These results serve to show that the results are various and that in such cases the use of indicators is very limited. Parallel experiments, following the course of the titration electrometrically, showed that the presence of the colloid depresses the concentration of hydrogen ions in acid solutions and of hydroxyl ions in alkaline solutions. The electrometric titration curve of hydrochloric acid and sodium hydroxide in the presence of 3 per cent. of gelatin or gum arabic resembles that of a weak acid and a weak base. In addition to the buffering effect of the colloid, it is considered that the degree of dispersion of the indicator partly accounts for the effect. Excepting solutions of methyl-orange and alkaline solutions of phenolphthalein, rosolic acid, and Congo-red, the indicators were found to be colloidally disperse.

COLLOID PRINCIPLES

The adsorption of dyes by colloids produced in volumetric reactions is sometimes utilized for the determination of the end-point. Fluorescein in solution gives the well-known yellow colour with green fluorescence, but when adsorbed on a white substance such as silver, chloride imparts to it an orange colour. If fluorescein is added to a silver chloride sol or suspension in the presence of excess of silver nitrate, as in a titration, the positively charged sol adsorbs the fluorescein and is coloured thereby. If the sol is in presence of an excess of sodium chloride it is negatively charged and the fluorescein is not appreciably adsorbed. Fluorescein can therefore be used as an adsorption indicator to determine the end-point. In practice, the concentration of the fluorescein is important.

The principles of colloid chemistry also enter into gravimetric analysis, particularly in respect of the precipitation of substances in such a form that thay can be filtered and washed without trouble and in a reasonably short time. The adsorption of the other substances by the precipitate is also an important matter.

Definitely crystalline precipitates are desirable for their ease of filtration and the procedure followed in routine analysis aims at producing a sufficiently large size of particle. Extremely small particles may go through the filter paper, but that is not the only source of inconvenience ; the greatest trouble arises with highly hydrated substances which are precipitated in a highly dispersed state. The primary particles then adhere to form a slimy gel of the type which is characteristic of precipitates of the hydroxides of aluminium and iron. The primary particles are small enough to pass through the filter paper, but generally do not do so because they are retained by electrical forces and because of their adherence to each other. The result is, however, that they block the pores of the filter and filtration and washing become very tedious and at times impossible. The application of suction merely serves to increase the blockage.

The particle size is determined by the rules laid down by von Weimarn (p. 9) and will be smallest when the solubility of the precipitate is extremely low and the concentration of the reacting solutions high. Another effect is that when the solubility is excessively small there will be practically no " ripening " of the precipitate or growth of large particles at the expense of the smaller. The solubility of ferric hydroxides is extremely small even compared with that of barium sulphate, and conditions are therefore right for the production of a gelatinous precipitate.

The concentration of the reagents, rate of addition, temperature of mixing, and presence of other salts, with other conditions which may appear at first sight to be mere details are therefore important in order to secure reproducible results in quantitative analysis. The secondary state of aggregation of the precipitate is affected considerably by other electrolytes present in the solutions, but it is not always possible to use this fact to advantage, because they are generally adsorbed by the precipitate. Acids and ammonium salts are favourite additional electrolytes because they can be removed by ignition of the precipitate.

An interesting fact in connexion with the coarsening of a slimy precipitate is encountered in a method of determining aluminium. The aluminium solution is neutralized with a few drops of ammonia and sodium thiosulphate solution is then added. The precipitate consists of a mixture of aluminium hydroxide and free sulphur, the latter being removed during the subsequent ignition. The precipitate, however, is much more granular than that produced by ammonia. This is because aluminium hydroxide is a positively charged colloid, whilst the colloidal sulphur produced simultaneously is negatively charged. The two colloids therefore mutually discharge and adsorb each other, giving a precipitate which can be filtered and washed more satisfactorily.

Adsorption is much greater the smaller the particle size of the precipitate and forms one of the most frequent sources of error in gravimetric analysis. Adsorption is generally less at higher temperatures and hence the reason for mixing hot rather than cold solutions. Errors due to adsorption may be either positive or negative. For example, in the determination of sulphuric acid by adding barium chloride adsorption of the barium salt occurs unless special precautions are used, the precipitate weight being increased thereby. In practice, the solution is kept hot and the barium chloride solution is added drop by drop so that at no time is the precipitate in the presence of a considerable concentration of barium ions. On the other hand, in the determination of barium chloride by means of sulphuric acid the barium chloride, which is the substance most easily adsorbed, is present from the start and there is no ready means of preventing its adsorption by the first particles of barium sulphate produced.

The adsorption error is negative, because the molecular weight of barium chloride is less than that of barium sulphate. In this case the error is obviated by treating the precipitate with concentrated sulphuric acid in the crucible and removing the excess by heating. The adsorption of foreign matter from the solution leads to a positive error. The importance of adsorption in these processes cannot be over-estimated and it has been stated that no case is known where barium sulphate is precipitated in a pure form. It is also necessary not to lose sight of the fact that the solvent is also adsorbed and may in certain cases be completely removed with difficulty. Precipitates of barium sulphate dried in the air were found to contain water generally in a direct ratio to the amount of adsorbed salt.

Dyestuffs. The dyestuffs include an enormous number of substances having very different properties, many of them being definitely basic or acidic and others readily hydrolysed. Their degree of dispersion is sometimes definitely molecular, sometimes definitely colloidal, and often in the transitional stage which is difficult to define. Most of the dyes are characterized by high molecular weight, so that molecular and colloidal dispersion are not incompatible, and on the whole they belong to a class which can be regarded as intermediate between colloids and crystalloids.

Recently, Nistler (*Kolloidchem. Beih.*, 1930, 31, 1) has applied a micro-method for the determination of diffusion coefficients to the examination of the dispersity of dyes. The dispersity of a given dye was found always to lie between two fairly close limiting values. Many freshly prepared solutions of dyes showed a rapid alteration in dispersity, reaching a stable state after some hours or days, which might persist or be followed by a further slow change in the direction of increased dispersity (aurantia, erythrosin, alkali-blue, methyl-green) or coarsening (eosin A, uranin). This peculiarity was shown especially with aurantia, the dispersity of which changed more during the first hour after preparation than during the following year. The effect is most noticeable in the more concentrated solutions. Safranine, acid fuchsin, Magdala-red, and other dyes showed no such ageing effect. No general statement can be made about the variation of dispersity with concentration, since this varies from dye to dye, but in many instances, and particularly with the more coarsely dispersed dyes, the dispersity remains constant. The particle size of the majority of dyes lies between 0·6 and 3·0$\mu\mu$ over the concentration range 0·1 to 0·001 per cent.

Many of them produce a marked lowering of the surface tension of water and it is therefore to be expected that they will be positively adsorbed. It is not at present possible to give a complete colloid-chemical theory of dyeing. In fact, the results obtained appear to be contradictory to a considerable degree. This is because dyeing has been studied more from a technical than a purely scientific point of view and the technical preparations do not always consist of pure substances. The practical importance of the problem is greater than the theoretical interest and the enormous amount of different observations and empirical methods have not been sorted out into a consistent whole.

Not only is the dyestuff colloidal, but the wool, silk, or other substance being dyed is also a colloid. It may be expected that the mutual discharge and adsorption of oppositely charged colloids plays an important part. The chief point at issue is whether the colouring of the fibre by a dye is a chemical process, a process of solid solution, or purely a case of adsorption. There are several investigations on record where the taking up of the dye is in accordance with the ordinary adsorption isotherm, but this certainly does not fit all cases. It is very difficult to generalize from a large amount of disjointed data, but the general impression received is that the dyeing process may be considered as one of adsorption upon which, in many instances, solvent and chemical action are superimposed.

Dyes which are not fast when used alone can be fixed by the use of a *mordant*. Mordants are not necessarily colloids, but produce colloids in the fibres. Typical examples are salts of aluminium, chromium, and iron, stannic chloride, tannin, etc. Aluminium sulphate solution is partly hydrolysed and when brought into contact with cotton the positively charged sol of aluminium hydroxide is fixed by the negatively charged cotton. The gels thus fixed can react with the colouring matter of the dye-bath, either through adsorption or by other means.

Leather. Both hide and leather are colloidal substances. The hide substance is mainly a mixture of condensation products of amino acids, the cell walls and their contents being similar in composition, although the latter are more soluble. There are essentially three steps in tanning : (1) Rehydration of the partially

COLLOID PRINCIPLES

dried hide; (2) removal of hair by the action of alkalis or alkaline sulphides; and (3) precipitation of the hide substance by the action of tanning agents.

The hide swells in water, but more rapidly in dilute acids or alkalis. In practice an alkali is used for the production of soft thin leather and dilute sulphuric acid for the production of heavy leather for belting or for the soles of shoes. The swelling must be stopped before degradation of the material occurs.

The function of the tanning agent, which consists of tannin or vegetable extracts, or salts of iron, aluminium, or chromium, is to form an insoluble precipitate with the hide substance, which swells very little. The tanning agents are either colloids or (as in the case of the inorganic salts mentioned) readily produce colloids, and almost certainly the action is one of mutual precipitation of colloids of opposite sign. Adsorption undoubtedly plays an important part in the first stages of tanning, and as the hide becomes more and more leather-like the amount of tanning agent which can be extracted by water diminishes.

Among the mineral tanning agents, chromium salts are the most efficient and ferric salts the least. The reason generally given is that ferric salts are so easily hydrolysed that they do not easily penetrate the substance of the cell.

Rubber. Rubber is obtained from the milk-like sap of certain trees and shrubs. The sap or latex is a colloidal system of globules of caoutchouc in an aqueous liquid. Caoutchouc is a polymer of C_5H_8, the molecular weight of which is unknown. The following table gives a typical analysis of caoutchouc.

Water	56·9	Resins	4·16
Caoutchouc	36·53	Protein and minerals	2·88

The particles of caoutchouc are in Brownian motion and their size varies over wide limits.

The separation of the disperse phase for the preparation of rubber is an industrial colloid-chemical problem. Heating and smoking, boiling, evaporation to cakes, dilution with water, and addition of precipitating agents, such as acids, alum, etc., are all employed on rubber plantations. The most efficacious treatment depends largely on the source and composition of the latex, the process being complicated by the protein content, which acts as a protective colloid to the caoutchouc. The effect of the boiling process is supposed to be due to the coagulation of the protective coating of protein at this temperature, exposing the globules of caoutchouc to coagulation by the electrolytes normally present in the latex. Experiments have shown that the effect of precipitating agents on latex is similar to that on proteins, and it seems that the protein film on the outside of the globules must be destroyed before an elastic mass can be formed.

In any case, the final effect is a coagulum of a very soft and sticky substance known as caoutchouc, which has to be vulcanized before it can be used for most purposes.

The process of vulcanization consists of the addition of sulphur in some form and has the effect of causing the mass to lose its adhesiveness and to be rendered insoluble in any liquid that does not permanently destroy it and made more resistant to oxidation. The elasticity of the product may be varied over a wide range, also.

In cold vulcanization the material is passed through a solution of sulphur chloride in carbon disulphide or carbon tetrachloride, the excess being afterwards neutralized with ammonia. Rubber goods prepared in this way are very soft, but tend to " perish " easily.

Hot vulcanization is more usual, the material being heated with sulphur along with an accelerator. Litharge is often used as an accelerator for black goods and lime or magnesia for white goods. For soft rubber, the amount of sulphur taken up varies from 2–10 per cent. For the production of hard rubber or vulcanite as much as 30 or 35 per cent. of sulphur may be added.

It is not certain whether the sulphur is chemically combined or adsorbed. Against the chemical theory is the fact that all the sulphur cannot be removed by a solvent such as acetone. Usually about 3 per cent. becomes permanently fixed. Possibly both adsorption and chemical interaction occur.

Milk. Milk contains three disperse phases—fat, casein, and albumin. The proteins, casein and albumin, act as a protective colloid for the fat particles and probably form an envelope round them, which prevents their coalescence.

When milk is left to stand, the fat particles with the casein rise to the top as cream. The separation may also be effected by centrifuging. The fat may be freed from the casein by severe mechanical agitation—churning—but only with difficulty unless

COLLOID PRINCIPLES

the cream has become sour. Small amounts of acid precipitate the casein and the unprotected fat (or butter) then comes out readily. In the making of butter from soured cream, the lactic acid produced during the souring destroys the protective action of the casein by precipitating it.

The homogenization of milk is a process for increasing the degree of dispersion of the fat and is accomplished by forcing the milk at about $85°$ between iron plates under a pressure of 250 atmospheres. The resulting more highly dispersed colloidal system is more stable, for the cream does not rise to the top on standing, nor can it be separated by centrifuging.

Wool. Textile materials are of colloidal nature and a knowledge of the behaviour of colloids should prove to have far-reaching results in the industries connected with textiles. Of recent years there have arisen in this country several Research Associations which have had the object of investigating the scientific aspects of the industries they represent. The first problems awaiting solution have been connected with the mechanical, physical, and purely chemical properties of the substance and the colloid-chemical point of view has scarcely been developed yet. It is not possible to give an indication of the results which have been achieved in each of these industries, but some recent developments in our knowledge of wool will serve to show the direction in which experiment and thought are travelling.

The wool substance, apart from grease and moisture, is almost entirely protein in nature and is one of the keratins—a group of proteins which are characterized by their insolubility and comparative resistance to enzyme hydrolysis as well as by a high content of sulphur. The constitution of wool keratin is unknown so far, but by degradation a number of amino acids are obtained from it. Cystine and glutamic acid are the chief products, closely followed by leucine, arginine, and histidine. The cystine is particularly important, since it is the only one among the products which contains sulphur, and the sulphur content has been found to have an important influence on the spinning quality of the wool. The great preponderance of basic amino acids confers a decidedly basic character on wool.

Wool grease consists mainly of esters of stearic, palmitic, and cerotic acids with the higher alcohols and is used in the manufacture of pharmaceutical preparations, soap, candles, lubricants, printing-ink, shoe-polishes and varnishes.

Wool has many colloidal characteristics. It is hygroscopic and has a considerable heat of wetting; there is also evidence that it is porous, probably having pores of a few $\mu\mu$ diameter. When the fibre is stretched the pores are distorted and there is an increase in the amount of water taken up. Wool swells in water and aqueous solutions and in many organic media, the cross-sectional area of the fibre increasing, whilst scarcely any change is noticeable in the length of the fibre. Initially elliptical fibres also tend to assume a more circular contour during the swelling. An investigation of the effect of hydrogen-ion concentration on swelling has shown a minimum swelling at $p_H 0 \cdot 1$, a maximum at $p_H 1 \cdot 25$ and another flat minimum at $p_H 4 \cdot 0$ to $4 \cdot 5$. Only in the case of skin has a minimum similar to that at $p_H 0 \cdot 1$ been observed.

The elastic properties of wool are consistent with those of a gel and it has been suggested that the wool fibre consists of two gels arranged in parallel. One of these consists of an elastic cell enclosing a fibrillar structure, which is not in equilibrium with a viscous phase. According to this view, the wool possesses three phases: an elastic wall, an enclosed fibrillar structure, and a viscous medium of gelatinous character included in its interstices. On the whole, these views receive support from the X-ray spectrographic investigation of the fibre.

An excellent review of the position has been written by Barker ("Wool: A Study of the Fibre," H.M. Stationery Office, 1929).

Water Purification. The colloids occurring in natural waters are mainly clay, organic colloidal matter, and bacteria. Bacterial cells in suspension have the typical properties of colloids; they are negatively charged, migrating towards the anode when an electric field is applied, and are precipitated by cations, especially those of high valency; on the whole, however, they are more stable than typical suspensoids, and alkali salts have little or no effect. The removal of bacteria, and particularly those which are responsible for diseases, is one of the major problems of water purification.

The organic colloidal matter is derived normally from decomposing vegetable matter. Water taken from a swamp is generally coloured, the colouring matter being mainly colloidal and as a rule positively charged.

In addition, water may be quite turbid with particles of clay or silt. These particles are negatively charged and therefore neutralize and coagulate the organic colloids. For this reason it is unusual to find a water which is both turbid and coloured, and if a coloured water from a swamp enters a stream containing colloidal clay, precipitation occurs. The formation of deltas at the mouths of rivers is an example of precipitation by the electrolytes of sea-water. The fine particles carried down by the river are precipitated on encountering the salt water and form muddy tracts, which would silt up the channel, were it not kept clear by dredging.

A modern and efficient method of water purification makes direct use of colloid principles. Aluminium or ferric sulphate is added to the water along with soda ash, causing the precipitation of the positively charged colloidal hydroxide of the metal. The water is then passed through a basin in which the hydroxide is coagulated and finally at a rapid rate through a bed of sand. Practically all the turbidity and colour and a very large percentage of the bacteria are removed, for not only does the positive colloid precipitate the negatively charged particles, but it also adsorbs colouring and other matters in true solution.

Sewage Disposal. The objectionable matter in sewage consists mainly of nitrogenous organic material of colloidal nature, which is negatively charged ; its disposal is therefore a problem of colloids, involving the precipitation of this material from the water. The technique of the processes employed is outside the scope of this book, but the principle involved is the precipitation of the colloidal matter in the sewage by contact with sand and crushed stone and, even more important, by contact with a gelatinous, slimy growth of micro-organisms, which forms on the surface of the stone. This growth is the essential feature of the trickling filter process.

A recent method for sewage purification is known as the " activated sludge process." If sewage is aerated for several weeks the particles settle, the water clarifies, and in time nitrification to harmless products takes place. This process is inconveniently slow, but it has been found that if some of the deposit from the settled sewage (or " activated sludge ") is added to a fresh sample of sewage, which is again aerated, the time of purification is greatly reduced. By repetition of the process, satisfactory results can be obtained in a few hours.

No satisfactory explanation of the activated sludge process appears to exist. The mutual precipitation of oppositely charged colloids is excluded because both the sewage colloid and the activated sludge are negatively charged. The sewage colloids have an isoelectric point at $p_H 4\cdot 6$ and Baly has suggested (*J. Soc. Chem. Ind.*, 1931, **50**, 23T) that if the process were operated at $p_H 5\cdot 8$ to $6\cdot 0$, where the colloids would be positively charged, an even greater efficiency would be attained.

Photography. The influence of light on a photographic plate and particularly the formation of the latent image has received much study of recent years, and although complete agreement has not been reached there is much to show that the problem is one to be studied by colloid-chemical methods.

A silver halide dispersed in a gelatin gel constitutes a photographic " emulsion." The colloidal silver halide may be produced readily by mixing gelatin solutions containing respectively dilute silver nitrate and a dilute alkali halide, but the sols so produced have a slow reactivity in the high state of dispersion. For rapid plates the film had to undergo a process of " ripening," whereby an increase in particle size of the silver halide occurs.

On exposure to light for a short time the latent image is produced at affected spots and on subsequent development with a reducing agent reduction to metallic silver occurs at these places. The unaltered silver halide may afterwards be dissolved out with sodium thiosulphate solution or " hypo." Modern investigations show that the exposed plate contains grains of silver halide in which minute nuclei of silver are produced as a result of the exposure, and it is only such grains already containing a nucleus of silver which are reduced to the metal when the plate is subsequently developed. The researches of Svedberg (*Phot. J.*, 1922, **62**, 183) have brought out very clearly that one silver nucleus in a grain of silver bromide makes the whole grain capable of subsequent development, whilst a grain not containing a nucleus remains undeveloped unless it is in contact with a grain containing such a nucleus.

The problem why certain parts of the grain are better adapted than others for the formation of the nucleus has received some attention and the impression gained is that the presence of catalysts favours the reduction at specific points.

Sheppard (*Phot. J.*, 1925, **65**, 380) has shown that certain decomposition products present in the gelatin favour the sensitiveness of the photographic film and it is probable that these catalysts are located at certain spots on the grain.

AUTHOR INDEX

Adam, N. K., 139
Amberger, C., 115
Angelescu, E, 181
Applebey, M. P., 204

Bachmann, W., 196
Baly, E. C. C., 262
Banerji, S. N., 197
Barker, S. G., 236, 260
Bary, P., 87, 231
Bechhold, H., 75, 76, 113, 219
Benson, Miss C., 122
Berzelius, J. J., 3, 188
Billitzer, J., 113
Biltz, W., 106
Bolam, T. R., 237
Bradford, S. C., 232, 239
Bredig, G., 17, 19, 68, 191
Brintzinger, H., 252
Brown, R., 39, 40
Butschli, O., 207
Buxton, 103
Buzágh, A. von, 45, 100, 134, 213, 214, 215

Chakravarti, M. N., 117
Chatterji, 233, 237
Chaudhury, 99
Clayton, W., 149
Cofman, V., 217
Copisarow, M., 207
Cottrell, 247

Dennett, J. H., 244
Dhar, N. R., 99, 101, 108, 110, 117, 233, 237
Donnan, F. G., 150
Draper, H. D., 276
Dreaper, W. P., 230

Eggert, J., 173
Einstein, A., 40, 41, 146, 155, 156, 157, 163
Eirich, F., 66

Elissafoff, G. von, 59
Ellis, R., 150
Evans, U. R., 126

Fairbrother, F., 185
Faraday, M., 3, 21, 23
Ferguson, J., 204
Fick, 222
Finkel, P., 276
Fischer, M. H., 190
Foerster, F., 250
Foulk, C. W., 159, 160
Frenkel, G., 181
Freundlich, H., 46, 53, 57, 85, 92, 93, 94, 95, 108, 128, 147, 165, 198, 240
Friend, J. N., 244
Fuchs, W., 66

Galecki, A., 242
Gallay, W., 192
Ganguly, P. B., 107
Ghosh, S., 99, 110, 197
Gibbs, W., 126
Gibbs, W. E., 243
Gortner, R. A., 164
Gouy, G., 57
Graham, T., 3, 4, 5, 48, 71, 188, 204, 219
Gutbier, A., 23, 252

Haller, W., 134
Handowsky, 175
Hardy, W. D., 53, 83, 90, 91, 93, 96, 97, 136
Harkins, D., 62, 138, 141
Hatschek, E., 157, 192, 198, 223, 229, 232
Haurowitz, F., 191
Hedges, E. S., 116, 125, 126, 131, 224, 228, 230, 231, 233, 234, 236, 237, 238, 239, 240, 241, 242, 244, 245

INDEX

Helmholtz, H. von, 57
Henley, Miss R. V., 116, 228, 230, 233, 234, 237
Heringa, G. C., 203
Heymann, E., 118, 194
Hildebrand, J. H., 276
Hoffman, W. F., 165
Hofmeister, 164, 171
Holker, J., 104, 105
Hooker, M. O., 190
Humphry, R. H., 192

Jacobsohn, K., 47
Jakovlava, 102
Jane, R. S., 192
Jermolenko, N., 216
Jong, H. G. B. de, 186, 196, 198

Katz, J. R., 206
Kestenbaum, P. P., 216
Köhler, R., 214, 216
Koller-Aeby, H., 242
Kraemer, E. O., 152
Krestinskaja, V. N., 102
Kross, W., 147
Kruyt, H. R., 99, 156, 162, 186
Krzeczkovska, I., 242
Kuhn, A., 205

Lal, P., 107
Langmuir, I., 62, 123, 138
Laplace, 121
Lea, Carey, 24
Lesche, E., 118
Liander, H., 243
Liepatov, S., 205
Liesegang, R. E., 223, 226, 229, 230, 232, 236, 239, 240
Lillie, R. S., 170
Linder, S. E., 51, 90
Lloyd, D. J., 217
Loebmann, S., 108
Lohfert, H., 37
Lottermoser, A., 182

McBain, J. W., 157, 186, 187
Mackenzie, Miss M. R., 237
Mark, H., 50
Mastin, H., 185
Meyer, K. H., 160, 179
Mircescu, J., 181
Moltschanova, O. S., 102
Mukherji, S. N., 100

Myers, J. E., 131, 231, 234, 242, 244

Neale, S. M., 183
Nernst, W., 60
Nistler, A., 255
Norris, Miss M. H., 236

Oakley, H. B., 247
Odén, S., 24, 34, 35, 45, 90, 136, 146, 147
Ostwald, Wi., 232, 233, 239
Ostwald, Wo., 7, 8, 9, 10, 35, 45, 74, 75, 85, 86, 134, 136, 156, 181, 207, 212, 213, 214, 215, 216, 233, 235, 239
Ott, A., 182

Paal, C., 115
Pauli, W., 66, 68, 175
Pavlov, P. N., 183
Pennycuick, S. W., 68
Perrin, J., 33, 34, 42, 43, 85
Pickering, S. V., 150
Picton, H., 51, 90
Poiseuille, 156, 157
Popp, K., 228
Posnjak, 200
Powis, F., 150
Prakash, S., 108
Prausnitz, P. H., 191
Pringsheim, N., 223
Puri, A. N., 249

Quast, A., 35

Rabinovitsch, M., 193
Rayleigh (Lord), 37, 138
Regener, 15
Reid, B. M., 108
Reinke, 200
Reitstötter, J., 173
Ròdiger, W., 215
Rona, P., 57

Schade, H., 37
Schalnikov, A., 191
Schroeder, J. von, 199
Schulze, H., 90, 91, 93, 96, 97, 136
Sen, K. C., 105, 237
Sheppard, S. E., 263
Sibi, M., 210

INDEX

Siedentopf, H., 7, 8, 76
Simon, A. L., 223
Sinclair, W. B., 165
Sladek, I., 216
Smith, C. G., 212
Smoluchowski, M. von, 40, 41, 99
Sörensen, S. P. L., 178, 216
Spring, W., 125, 187
Stamm, A. J., 152, 184
Staudinger, H., 179
Steinbach, W., 214
Stokes, 89
Susich, G. von, 50
Svedberg, T., 14, 19, 20, 41, 44, 45, 84, 174, 177, 178, 191, 262

Tammann, G., 29
Teague, 103
Thomas, P., 210
Thorne, P. C. L., 212

Traube, J., 20
Trillat, J. J., 49
Tyndall, J., 36

Van der Waals, J. D., 121

Wackenroder, 3
Wagner, H., 213
Weimarn, N. von, 13
Weimarn, P. P. von, 9, 10, 21, 22, 27, 67, 183, 202, 208, 210, 253
Whitby, G. S., 192
Williams, 237
Wintgen, R., 173

Zeigler, W., 219
Zocher, H., 47
Zsigmondy, R., 7, 8, 12, 21, 22, 34, 39, 40, 45, 47, 60, 61, 67, 76, 110, 111, 114, 207
Zyvotınski, P. B., 193

SUBJECT INDEX

Absorption, 120
Acclimatization, 92, 101
Acidoid, 249
Adherence, 100, 135
Adhesion, 142
— probability of, 98
Adsorption, 63, 95, 119, 214
— isotherm, 128
Aerogels, 243
Aerosols, 243
Agar, 184
Agate, 188, 226, 240
Ageing, 47, 117
Albumins, 112, 169, 174
Albumoses, 170, 173
Alkali metal sols, 91, 191
Aluminium, determination of, 254
Amalgams, colloidal, 193
Amicrons, 8
Amino-acids, 168
Amorphous state, 48, 210
Amylopectin, 181
Amylose, 180
Analysis, 251
Angle of tilt, 100
Anomalous structures, 229, 230
Anti-foams, 160
Arabinosazone, 210
Arsenious sulphide sol, 27, 92, 94, 97, 102, 107, 236
Avogadro constant, 42

Bacteria, 260
Banded structures, 226
Barium sulphate gels, 10, 209
— — precipitates, 9, 254
Benzosols, 12, 26
Biological structures, 225, 226, 240
Blood, 176
— pressure, 171
Boiling-point, rise of, 32
Boundary phenomena, 121
Breathing, 246

Brownian movement, 39
Bubbles, shape of, 198
Butter, 259

Calcium acetate gel, 212
Camphorylphenylthiosemicarbazide, 192
Capillaries, reactions in, 225, 230, 234, 236
Carbohydrates, 179
Casein, 177, 258
Catalase, 242
Catalysis, 131, 242
Cataphoresis, 51, 78, 191
Caoutchouc, 50, 203, 206, 257
Cellular structures, 208, 226
Cellulose, 49, 182, 202
— acetate, 49, 192
— nitrate, 49, 192, 197
Centrifuging, 33, 178
Ceramics, 248
Charcoal, as adsorbent, 119, 128, 132
Charge, determination of, 55, 78
— distribution of, 196
— origin of, 58, 191
Chitin, 202
Classification of colloids, 11, 82
Clays, 248
Clouds, 13
Coagels, 195
Coagulation, 61, 90, 106, 107, 108, 162, 175
— fractional, 34
— periodic, 236
— velocity of, 99
Coalescence, 109
Cohesion, 141
Collagen, 170, 203
Collargol, 115
Collision, probability of, 98
Collodion, 72, 74
Colloid mill, 15

INDEX

Colour of sols, 22, 47
Concentration, effect on stability, 99
— function, 129
Condensation, process of, 21
Condensers, sub-stage, 78
Conductivity of sols, 56
Continuous phase, 7
Cotton, 182
Cream, 258
Critical concentrations, 240
Crystalline particles, 48, 207
Crystallization, 29, 210, 223
Cuprammonium solutions, 182
Curds, soap, 187
Cuvette, 77
Cystine, 259

Deformation of gels, 198, 199, 206
Dehydration, 162
Denaturation of proteins, 175
Desorption, 131
Detergent action, 187
Dextrin, 179
Dialysis, 3, 71
Dielectric constant, 59, 134
Diffuse double layer, 57
Diffusion, 3, 208, 219
— coefficient, 35
— waves, 233, 235, 239
Dipole moment, 132
Disc-and-ring structures, 230
Discontinuous phase, 7
Discrete structures, 226
Dispargen, 115
Dispergation, 183, 202
Disperse phase, 7
Dispersion, degree of, 8
— medium, 7
— process of, 14, 183, 202
Dispersoids, 7, 10
Dissociation methods, 25
Double decomposition, 27, 225
Dust, 15
Dyes, 35, 255

Egg-white, 112, 170, 174, 176
Elasticity of gels, 206, 260
Elastogels, 213
Electric charge, 55, 58, 78, 191, 197
— double layer, 57, 97
Electrical properties, 51
Electro-adsorption, 135

Electrodialysis, 73
Electrodispersion, 17
Electrokinetic potential, 58
Electrolytes, action of, 53, 90, 202
— adsorption of, 136
— colloidal, 56, 109
Electrostatic potential, 58
Electro-viscous effect, 157
Emulsions, 16, 144
Emulsoid colloids, 84
Enzymes, 170, 242
Evolution zone, 87
Exchange adsorption, 133, 137, 249

Ferments, 242
Ferric hydroxide sol, 16, 28, 94, 102, 107
Fibres, structure of, 182, 184, 202, 203, 208, 211, 236, 260
Fibrin, 169, 202
Fibroin, 170, 202
Fick's law, 222
Films, structure of, 138
Filter paper, charge on, 74, 136
Filtration, 135, 251
Flocculation, 91
Fluorescence, 36
Foams, 11, 159, 213
Freezing-point depression, 32
Froth, 159
Fumes, 247

Gamboge, 42
Gelatin, 171, 207, 216, 219
Gelatinous precipitates, 195, 210, 253
Gelation, 195, 207
Gelatose, 172
Gels, properties of, 195
— reactions in, 116, 219
— structure of, 207
Geranin gel, 205
Gibbs' formula, 126
Glands, secretions of, 205
Glass, 21, 49, 208
Globulins, 113, 169, 176
Glue, 170
Glutin, 171
Gold number, 111
— sols, 3, 18, 21, 34, 66, 115
Graphite, colloidal, 250
Gravity, effect of, 43, 89
Growth of plants, 200

INDEX

Hæmocyanin, 178
Hæmoglobin, 31, 176, 178
Hair, 170, 257
Heart-beat, 246
Heterogeneity, 6, 38
Hide, 256
Hofmeister series, 164, 171, 202
Humic acid, 250
Hydration of ions, 165
— of particles, 145, 156, 163, 196, 201, 209
Hydrolysis, 28
Hydrophilic colloids, 85, 154, 168
Hydrophobic colloids, 85, 144
Hydrosols, 12
Hygroscopicity, 145
Hysteresis, 160, 185, 197

Ice, colloidal, 13
Ice-cream, 13
Indicators, colloidal, 253
— effect of colloids on, 252
Intermicellar liquid, 7
Intestines, 170, 171
Ionic micelle, 187
Ions, colloidal, 56, 161, 187
Irregular series, 102
Irreversible colloids, 83
Isodispersity, 33
Isoelectric point, 54, 166

Jellies, 195

Keratin, 170, 202, 259
Kinetic theory, 42, 43

Langmuir-Harkins theory, 138
Latex, 257
Lattice blocks, 20
Lead iodide rings, 235
— — sols, 117
Leather, 219, 256
Liesegang rings, 226
Light, influence of, 108, 229
Liquid-line corrosion, 125
Liquid-liquid boundary, 122
Liquid-solid boundary, 124
Liquid-vapour boundary, 121
Liquogels, 213
Lyophilic colloids, 85
Lyophobic colloids, 85
Lyosorption, 134

Lyospheres, 135, 213
Lyotropic series, 164, 171, 202
Lysalbic acid, 115

Magnesium hydroxide rings, 227
Mastic, 42
Mechanical dispersion, 15
Medicines, 2, 115, 171, 221
Melting of gels, 197
Membrane, dialysis, 3, 71
— semipermeable, 3
Mercuric sulphide sol, 28
Mercurosols, 193
Mercury sols, 18, 87, 191
Metal hydrosols, 66
Metals, dissolution of, 244
Micelle, 7, 187
Microns, 8
Milk, 149, 177, 258
Minerals, 12, 40
Mist, 11, 13
Mobility of particles, 56
Molecular weight, 31, 173, 177
Mordants, 256
Muscular contraction, 203

Negative sols, 52, 62
Nickel benzosol, 26
Night-blue, 131, 136, 154
Nitrocellulose, 74
Non-aqueous colloids, 170
Nuclear method, 34, 35
Nuclei, crystallization, 29, 34, 210, 222
Nucleoproteins, 169

Œdometer, 200
Oildag, 250
Oil emulsions, 149, 214, 248
Oil films, 139
Ointments, 2
Oligodynamic action, 220
Opacity, periodic, 105
Opal, 188
Opalescence, 6, 36
Optical properties, 36
Organosols, 12, 19, 190
Orientation, 62, 97, 138, 201
Osmotic pressure, 31, 42, 170
Ovomucoid, 113
Oxidation methods, 24
Oxyhæmoglobin, 176

INDEX

Palladium sols, 24, 115
Paper, 182
Paraboloid condenser, 78
Parchment, 3, 71
Particles, shape of, 37, 46
— size of, 7, 8, 9, 37, 44, 88, 224
Passivity of iron, 125, 240
Pastes, 208
Peat, 248
Pectographs, 231
Pepsin, 170
Peptides, 168
Peptization, 16, 61, 109, 137, 213, 215
Peptones, 170, 173
Periodic crystallization, 231
— opacity, 105
— reactions, 234, 244
— structures, 227
Permeability of membranes, 75
Phase ratio, 150
Phosphorus sol, 26
Photography, 219, 224, 262
Physiological effects, 115, 171, 205, 221, 245
Plastogels, 213
Platinum sols, 18, 68, 115
Poiseuille's law, 156
Polarity, 132, 142
Pollen, 39
Polydispersity, 33
Polymerization, 179
Polypeptides, 168
Polyphase systems, 11
Positive sols, 52, 62
Practical methods, 70
Precipitation, electrical, 247
— laws of, 9, 223
Pringsheim's rule, 223
Protalbic acid, 115
Protected sols, 114
Protective effect, 110, 113, 224
Proteins, 168
Purple of Cassius, 3, 21
Pyrosols, 193

Quartz particles, 100, 134

Radiations, influence of, 107
Reactions in gels, 116
Reactivity of colloids, 241
Reduction methods, 21
Refraction, double, 48, 206

Refractive index, 36, 39
Reversible colloids, 83
Rhythmic reactions, 234, 244
— structures, 226
Rigidity of gels, 206
Ring structures, 226
Ripening, 224, 253, 262
Rock salt, blue, 25
Rubber, 49, 192, 257
Ruby glass, 21

Saliva, 21
Salting-out, 163, 237
Scattering of light, 36
Schulze-Hardy rule, 91, 136
Schweizer's reagent, 182
Seaweed, 184, 200
Sedimentation, 44, 45, 134, 177
Selenium sols, 24
Sensitization, 109
Serum-albumin, 175
Setting of gels, 195
Sewage, 261
Shape of particles, 37, 46
Shrinkage, 203, 204
Silica gel, 204, 226
Silicic acid sols, 188
Silk, artificial, 184
— natural, 170, 202
Silver sols, 18, 23, 68, 115, 221
— chloride sols, 26, 63, 209
— chromate sols, 116
— — rings, 226
— iodide sols, 26, 63
Size of particles, 7, 8, 9, 37, 44, 88
Skin, 203, 260
Sky, colour of, 37
Sludge, activated, 261
Smokes, 11, 247
Soaps, 6, 151, 186, 190
Soils, 248
Solid-gas boundary, 122
— -phase rule, 213
— sols, 25
Sols, 4
Solubility, 9, 143, 216, 224
Solvation, 132, 135, 145, 157, 192, 201
Solvent, exchange of, 26
— influence on adsorption, 132
Sorption, 120
Space-lattice, 63, 123
Specific surface, 120

INDEX

Spiral structures, 203, 230, 236
Spongin, 170
Sprays, 15
Stability, 87, 89, 160
Stannic oxide sol, 16, 60
Starch, 180, 240
Starch-iodide, 181
Stirring, effects of, 48, 108, 198, 210, 212
Stokes' law, 45, 89
Stomach, 170
Streaming potential, 55
Structure of gels, 207
— particles, 48
Structures, cellular, 208, 226
— discrete, 226
— periodic, 227
Submicrons, 8
Sulphur sols, 24, 26, 148
Supersaturation, 9, 28, 116, 224, 236
Surface energy, 121
— phenomena, 119
— tension, 12, 121, 127, 146, 158
Suspensions, coarse, 8, 100, 214
Suspensoid colloids, 84
Swelling, 155, 199, 216
Syneresis, 185, 204

Tactosols, 47
Tanning, 207, 219, 256
Temperature, effect on stability, 108
Tendons, 203
Tetanus, 112
Textiles, 259
Thermodynamics, 217
Thixotropy, 198, 210
Titanium dioxide sols, 147
Titrations, 252

Tobacco juice, 21
Trypsin, 170
Turbidity, 92, 104, 261
Tyndall effect, 36

U-numbers, 110
Ultracentrifuge, 84, 178
Ultrafilters, 45, 73
Ultrafiltration, fractional, 35, 76
Ultramicrocrystals, 208
Ultramicroscope, 6, 38, 44, 76
Ultra-violet light, 108
Uniformity of particles, 33
Unimolecular layers, 124, 138
Urinary calculi, 240
Urine, 21

Valency rule, 91
Vanadium pentoxide sols, 47
Vapour pressure, 32
Viscogels, 213
Viscose, 184, 204
Viscosimeter, 80
Viscosity, 79, 86, 146, 155
Volume change in swelling, 201
Volume of particles, 147, 155
Vulcanization, 258

Water, as disperse phase, 13, 149
— purification, 260
— purity of, 22, 67, 70
— Tyndall light in, 38
Waterdag, 250
Wetting, 153, 260
Wool, 170, 236, 259

X-ray analysis, 48, 50, 63, 180, 204, 206
X-rays, influence of, 107
Xerogels, 213

CPSIA information can be obtained at www.ICGtesting.com
Printed in the USA
LVOW052348151212

311836LV00001B/200/A